Atmospheric Ultraviolet Remote Sensing

This is Volume 52 in the
INTERNATIONAL GEOPHYSICS SERIES

A series of monographs and textbooks
Edited by RENATA DMOWSKA and JAMES R. HOLTON

A complete list of the books in this series appears at the end of this volume.

Atmospheric Ultraviolet Remote Sensing

Robert E. Huffman

Phillips Laboratory
Hanscom AFB, Massachusetts

ACADEMIC PRESS, INC.
Harcourt Brace Jovanovich, Publishers
Boston   San Diego   New York   London   Sydney   Tokyo   Toronto

QC
871
.H84
1992

This book is printed on acid-free paper. ⊚

Copyright © 1992 by Academic Press, Inc.

All rights reserved
No part of this publication may be reproduced or
transmitted in any form or by any means, electronic
or mechanical, including photocopy, recording, or
any information storage and retrieval system, without
permission in writing from the publisher.

ACADEMIC PRESS, INC.
1250 Sixth Avenue, San Diego, CA 92101-4311

United Kingdom Edition published by
ACADEMIC PRESS LIMITED
24–28 Oval Road, London NW1 7DX

Library of Congress Cataloging-in-Publication Data:

Huffman, Robert E., date.
    Atmospheric ultraviolet remote sensing / Robert E. Huffman.
      p. cm. — (International geophysics series: v. 52)
    Includes bibliographical references and index.
    ISBN 0-12-360390-0
    1. Atmosphere, Upper—Remote sensing. 2. Ultraviolet radiation—
Remote sensing. I. Title. II. Series.
QC871.H84   1992
551.5' 14—dc20

                                                                92-26268
                                                                      CIP

Printed in the United States of America
92 93 94 95 BC 9 8 7 6 5 4 3 2 1

Contents

	Preface	ix
1	**Introduction**	**1**
	1.1 Ultraviolet applications	2
	1.2 Scope of this book	3
	1.3 Intended audience	4
	1.4 General references	4
	1.5 References	6
2	**The UV—What, Where, and Why**	**7**
	2.1 The ultraviolet defined	7
	2.2 Global ultraviolet	11
	2.3 Atmospheric ultraviolet	12
	2.4 Why use UV?	17
3	**Radiometry**	**21**
	3.1 Photons, watts, and Rayleighs	21
	3.2 Radiance and irradiance	23
	3.3 Radiometric equations	25
	3.4 Spatial considerations	29
	3.5 Photon statistics	30
	3.6 References	33
4	**Sensors**	**35**
	4.1 Sensitivity	36
	4.2 Photometer	37
	4.3 Spectrometer	40
	4.4 UV imager	45

	4.5 Laboratory and space calibration	49
	4.6 References	53
5	**Space Operations**	**57**
	5.1 UV absorption considerations	57
	5.2 Satellites and orbits	62
	5.3 Integration and test	66
	5.4 Glows and plumes	68
	5.5 Major satellite programs	72
	5.6 References	73
6	**The Earth's Atmosphere**	**77**
	6.1 Atmospheric regions	78
	6.2 Atmospheric models	82
	6.3 References	84
7	**Solar Photoabsorption**	**87**
	7.1 Quiet sun flux values	88
	7.2 Solar flux variability	92
	7.3 Extinction and absorption	93
	7.4 Photodissociation and photoionization	97
	7.5 References	100
8	**Photon Cross Sections**	**103**
	8.1 Energy levels and equivalent wavelengths	104
	8.2 Absorption and ionization cross sections	104
	8.3 References	115
9	**Airglow**	**121**
	9.1 Excitation and measurement	121
	9.2 Day airglow	123
	9.3 Night airglow	127
	9.4 References	134
10	**Aurora**	**137**
	10.1 The auroral oval	138
	10.2 Auroral spectroscopy	140
	10.3 Auroral emission modeling	149
	10.4 References	150

CONTENTS

11 Scattering and Fluorescence — **155**
- 11.1 Rayleigh scattering — 155
- 11.2 Fluorescence — 160
- 11.3 Polar mesospheric clouds — 163
- 11.4 References — 167

12 Atmospheric Ultraviolet Backgrounds — **171**
- 12.1 Day backgrounds — 172
- 12.2 Night backgrounds — 181
- 12.3 Auroral backgrounds — 182
- 12.4 Clutter — 183
- 12.5 References — 186

13 Radiance and Transmission Codes — **189**
- 13.1 LOWTRAN and similar user codes — 190
- 13.2 AURIC UV code — 191
- 13.3 Validation and verification — 196
- 13.4 Related codes — 200
- 13.5 References — 201

14 Ozone and Lower Atmospheric Composition — **207**
- 14.1 Atmospheric ozone — 208
- 14.2 Local stratospheric ozone measurements — 213
- 14.3 Global stratospheric ozone measurements — 214
- 14.4 Minor species — 218
- 14.5 References — 219

15 Upper Atmospheric Composition and Density — **227**
- 15.1 Absorption methods — 228
- 15.2 Fluorescence and scattering methods — 232
- 15.3 Airglow methods — 236
- 15.4 Recommended methods — 243
- 15.5 Atmospheric density — 244
- 15.6 References — 245

16 Global Auroral Imaging — **253**
- 16.1 The aurora from space — 253
- 16.2 Satellite UV imagers — 255
- 16.3 Comparison with other methods — 267
- 16.4 Solar-terrestrial physics applications — 270
- 16.5 References — 270

17 Ionospheric Electron Density   **279**

17.1 Radiowaves and electron densities 280
17.2 Electron densities from UV radiance 283
17.3 UV solar flux approaches 291
17.4 Ionospheric irregularities 292
17.5 Global space weather systems 294
17.6 Concluding remarks 298
17.7 References 298

Index **307**

Preface

This book is an introduction to the use of the ultraviolet for remote sensing of the Earth's atmosphere. It covers the Earth's UV radiative environment, experimental techniques, and current applications. It is my intention to provide the information needed to "make a first approximation" concerning the use of the ultraviolet and to provide access through the literature for a more thorough study.

The ultraviolet has become more important to our life on Earth recently. Ultraviolet methods are used, both from space and from the ground, to monitor *ozone* in the atmosphere and its possible changes. The solar UV-B radiation reaching the Earth may well be increasing, bringing with it hazards to human health and the biosphere in general.

Not so well known is the use of ultraviolet methods to monitor the aurora and the ionosphere by global imaging from space. These methods will help make communications and radar operations more reliable through development of *global space weather systems*. Other applications include use of the airglow for atmospheric density and composition measurements and the development of atmospheric radiance and transmission codes as tools enabling ready assessments of proposed ultraviolet uses.

The cover of this book illustrates these new ultraviolet uses in the ionosphere and thermosphere. The sketch is based on ultraviolet images of the auroral oval we have obtained from the AIRS imager on the Polar BEAR satellite. Superimposed on the auroral image are the land mass outlines seen by a satellite at 1000 km altitude above the north geographic pole. The image is obtained in the 135.6 nm emission line of atomic oxygen.

PREFACE

There are few books specifically covering the ultraviolet, which is defined here as being from about 400 to 10 nanometers (4000 to 100 Å). An interdisciplinary approach is used to provide an introduction to the relevant atmospheric physics and chemistry, or aeronomy; the needed ultraviolet technology; and the current state of applications.

While the book emphasizes passive UV remote sensing from space satelites, it should be useful to those interested in related technical areas involving use of the ultraviolet for astronomy, laboratory spectroscopy, lidar applications, and related fields.

This book is based on more than thirty years in ultraviolet research and development at the Phillips Laboratory, Geophysics Directorate. Previous names for this laboratory have included the Air Force Geophysics Laboratory and Air Force Cambridge Research Laboratory. I have been involved with ultraviolet projects in laboratory spectroscopy; development of new light sources and calibration methods; measurement of missile plumes from sounding rockets; satellite and shuttle measurements of the airglow and aurora; auroral imaging from space; and critical reviews of phenomenology for UV radiance and transmission models.

This book began as a course on ultraviolet radiation given in association with three conferences on ultraviolet technology at SPIE meetings in 1986, 1988, and 1989. The material covered has been greatly expanded, and a more complete set of references is provided.

It is a pleasure to acknowledge the contributions of scientific collaborators at the Geophysics Directorate, including J. C. Larrabee, F. J. LeBlanc, F. P. DelGreco, C. G. Stergis, V. C. Baisley, R. W. Eastes, and L. A. Hall. I thank Marji Paulson for work on several illustrations and the research library staff for much assistance. The approval by the Geophysics Directorate management of this project is acknowledged. My son, Robert A. Huffman, has read and helped to edit the manuscript. The staff of Academic Press, including especially Robert Kaplan, Senior Editor, have been very helpful. Finally, many other people have been consulted about various aspects of the book, and their contribution is gratefully acknowledged.

This book is dedicated to my late wife, Jacquelin, whose support and encouragement were essential in the writing of this book.

Robert E. Huffman
Sudbury, Massachusetts

Chapter 1

Introduction

In pictures taken from space, the Earth appears as a globe of varying blues, greens, and whites traveling through the void. When these images were first obtained about twenty-five years ago, they helped give rise to the concept of a finite spaceship Earth, and a global viewpoint became easier to accept. At about the same time, the less dramatic but vital work of studying our planet in all its parts began to demonstrate that the activities of mankind were modifying spaceship Earth, mostly in undesirable ways. Concern with the global environment began.

The Earth can also be imaged from space by ultraviolet wavelengths. In different ultraviolet wavelengths, the emission may come from airglow, aurora, and scattering sources in the atmosphere. The image is therefore very different from the solid Earth features seen in a visible image. Ultraviolet images, appearing in quantity about ten years ago, have stimulated interest in using the UV to investigate and remotely sense one of the most fragile and variable parts of our environment: the stratospheric and ionospheric regions of the atmosphere.

One key area involves the measurement of stratospheric ozone, the ultraviolet flux reaching the Earth, and the long term changes in the two. Due to the nature of UV radiation and its interaction with the atmosphere, it is also involved in considerations of global warming. The apparently trivial, but ultimately large and global, changes in our atmosphere due to human activities such as the use of fossil fuels and fluorocarbons must be monitored in many ways. Ultraviolet techniques are among them.

Less well known is the use of the UV in the development of remote sensing methods for the ionosphere and aurora. In this case, the most immediate involvement with human life is in connection with radio propagation and its use for communications, radar, and navigation. Future monitoring of the ionosphere from space will improve the operations of these vital services.

This book deals with fundamentals, techniques, and applications of atmospheric ultraviolet remote sensing. The emphasis is on passive sensing of the Earth's atmosphere from space, but the atmospheric properties and experimental techniques discussed are important for active methods, such as lidar, as well. Passive methods involve use of naturally occurring emission from airglow, aurora, and scattering. In addition, passive sensing includes the occultation of UV sources, such as the sun by photoabsorption in the atmosphere, and the use of solar flux measurements in global atmospheric models. The altitude range covered extends from ground level to the magnetosphere, with most of the emphasis from the stratosphere through the thermosphere.

1.1 Ultraviolet applications

Some important applications areas for UV remote sensing include:

Stratospheric ozone The most important application of ultraviolet remote sensing at this time is its use in stratospheric ozone measurement. This research area is extremely active, with many space and ground programs. There is great concern about stratospheric ozone depletion and the resultant increase in solar ultraviolet radiation at ground level.

Global auroral imaging A recent development is the use of the ultraviolet for day and night global auroral imaging from space. The UV images of the auroral oval enable real time knowledge of the location and strength of the auroral zone as well as improved understanding of the particles and fields in this region.

Global space weather systems An important emerging development is the combination of auroral imaging, airglow, fluorescence, in-situ, and possibly other measurements from space together with ground-based measurements and the necessary models into a global space weather system. A system of this type would provide ionospheric electron densities and other information about the thermosphere and ionosphere needed to improve the operation of communications, navigation, and radar systems.

1.2 Scope of this book

The overall purpose of this book is to serve as an introduction to the use of the ultraviolet for remote sensing in, through, and of the atmosphere. In order to understand this subject, it is necessary to also have some understanding of both atmospheric geophysics and ultraviolet technology. This book seeks to combine these subjects with an interdisciplinary approach.

Technology transition is one goal of this book. Research over the last several decades has improved our knowledge of the atmosphere and its ultraviolet radiative environment to such an extent that applications useful to mankind are now possible. This book is planned to facilitate the transition of this technology from experimentation to societal use.

It is anticipated that this book will create interest in the use of UV methods and that it will serve as a source of ideas for further research and development.

This book seeks to describe and to explain, rather than to provide an exhaustive treatise in any phase of the subject. Fundamentals include the relevant physics and chemistry of the atmosphere, which is sometimes called **aeronomy**. Ultraviolet technology is discussed through descriptions of representative sensors and their use in space programs. The relationships of ultraviolet to infrared, visible, and x-ray methods are pointed out. Finally, the last six chapters describe applications areas. Detailed reference lists are given to lead the interested reader to the original literature.

Solar, astrophysical, and planetary studies are only included in this book where they are directly related to remote sensing of the Earth's atmosphere.

Throughout the book, the emphasis is on the use of the ultraviolet to solve the problems of people within our shared environment. All solutions are partial solutions, and, speaking philosophically, all sensing is remote. These applications to remote sensing are another small step to help understand and improve our world.

Some idea of the relationship between ultraviolet remote sensing and the major problems we face can be gained from the series entitled "State of the World," prepared yearly by the Worldwatch Institute. In particular, *Brown et al.*, 1988, emphasizes the importance of recent changes in the atmosphere and its chemistry, including air pollution, ozone depletion, and the buildup of greenhouse gases. In all of these topics, UV radiation measurements and techniques are involved in some way.

1.3 Intended audience

This book is addressed to the following types of people:

- Engineers and scientists using or considering the use of the ultraviolet for any sort of remote sensing.

- Advanced students specializing in atmospheric physics, atmospheric chemistry, aeronomy, ionospheric physics, etc.

- Program managers and others at all levels of management interested in development of new remote sensing capabilities.

- Research workers seeking an introduction, overview, or update of this field.

- Atmospheric modelers who wish an introduction to UV experimentation, and experimenters who wish an introduction to current UV atmospheric codes.

1.4 General references

Books that have been found valuable for many areas covered herein are as follows:

Brasseur, G. and S. Solomon, *Aeronomy of the Middle Atmosphere*, Reidel, 1984.

Chamberlain, J. W. and D. M. Hunten, *Theory of Planetary Atmospheres, An introduction to their physics and chemistry*, Academic Press, 1987.

Green, A. E. S., Editor, *The Middle Ultraviolet: Its science and technology*, Wiley, 1966.

Jursa, A. S., Scientific Editor, *Handbook of Geophysics and the Space Environment*, Air Force Geophysics Laboratory, Hanscom Air Force Base, Bedford, Massachusetts, 1985.

Rees, M. H., *Physics and Chemistry of the Upper Atmosphere*, Cambridge U. Press, 1989.

Samson, J. A. R., *Techniques of Vacuum Ultraviolet Spectroscopy*, Wiley, 1967.

1.4. GENERAL REFERENCES

Spiro, I. J. and M. Schlessinger, *Infrared Technology Fundamentals*, Dekker, 1989.

Zaidel', A. N. and E. Ya. Shreider, *Vacuum Ultraviolet Spectroscopy*, Ann Arbor-Humphrey Science Publishers, 1970 (translation by Z. Lerman, originally published in Russian in 1967).

Major journals covering the subjects in this book are:

- Journal of Geophysical Research (A: Space Physics)
- Journal of Geophysical Research (D: Atmospheres)
- Planetary and Space Science
- Applied Optics
- Reviews of Scientific Instruments
- Optical Engineering

Principal conference proceedings and abstracts used are as follows:

- Ultraviolet and Vacuum Ultraviolet Systems, *SPIE, 279*, W. R. Hunter, Editor, 1981.

- Ultraviolet Technology, I, II, III , *SPIE, 687, 932, 1158*, R. E. Huffman, Editor, 1986, 1988, 1989.

- The 9th International Conference on Vacuum Ultraviolet Radiation Physics, Proceedings edited by D. A. Shirley and G. Margaritondo, *Physica Scripta, T31*, 1990. Also see other proceedings of these valuable conferences begun by G. L. Weissler.

- Regular conferences and meetings of the American Geophysical Union and the Optical Society of America.

Some older references, in chronological order, useful in tracing the historical development of the ultraviolet in aeronomy include:

Mitra, S. K., *The Upper Atmosphere*, The Royal Asiatic Society of Bengal, 1 Park Street, Calcutta 16, 1947. For many years, this book was widely considered to be the first book to read on the upper atmosphere, as it was a comprehensive review of then current measurements and theory. It is still fascinating to explore.

Kuiper, G. P., Editor, *The Earth as a Planet*, U. Chicago Press, 1954. This collection of papers helped bring together scientists working in the laboratory and on field observations to provide a comprehensive description of the atmosphere.

Zelikoff, M., Editor, *The Threshold of Space, The Proceedings of the Conference on Chemical Aeronomy*, Geophysics Research Directorate, Air Force Cambridge Research Center (now Phillips Laboratory, Geophysics Directorate), June, 1956, Pergamon Press, 1957. This book provides a summary of developing knowledge of the atmosphere in the period just before satellite experimentation became possible.

Bates, D. R., Editor, *The Earth and Its Atmosphere*, Basic Books, 1957. This book is a popular account of the state of knowledge of the atmosphere during the International Geophysical Year (IGY), July 1, 1957, to December 31, 1958.

Ratcliffe, J.A., Editor, *Physics of the Upper Atmosphere*, Academic Press, 1960. The state of atmospheric knowledge immediately after the IGY is given by this book.

1.5 References

Brown, L. R. and staff of Worldwatch Institute, *State of the World 1988*, Norton, 1988, and continuing yearly volumes in this series.

Chapter 2

The UV—What, Where, and Why

Before detailed consideration of the fundamentals of UV remote sensing, it is necessary to describe *what* kinds of UV phenomena occur in our atmosphere, *where* the UV is in wavelength and altitude, and *why* the UV is important for atmospheric remote sensing.

2.1 The ultraviolet defined

Electromagnetic radiation at wavelengths shorter than the visible region and longer than the x-ray region is called ultraviolet radiation. The counterpart to the ultraviolet extending beyond the long wavelength end of the visible region is the infrared. The relationship of the ultraviolet to the electromagnetic spectrum is shown in Figure 2.1.

For this book, the ultraviolet is considered to extend from about 400 to about 10 nanometers; or 4000 to 100 Å; or 0.4 to 0.01 micrometers (commonly called microns). The photon energy range is from about 3 to 120 electron volts. These limits and the further subdivisions given herein are approximate and are not meant to be applied rigidly.

An attempt will be made to use the nanometer (nm) as the unit of choice for wavelength, although Ångstom (Å) , will occasionally appear. The nanometer as a unit of length in the ultraviolet is being used more frequently, especially by atmospheric chemists and in the longer wavelength part of the ultraviolet. The Ångstom is favored in the older literature and also by astronomers and many spectroscopists. The micrometer, or micron,

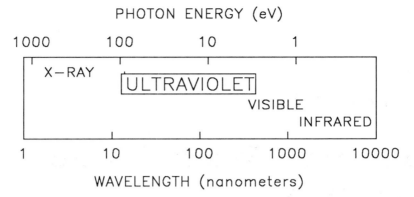

Figure 2.1: Ultraviolet in the electromagnetic spectrum

as a unit for the ultraviolet will be found generally among workers whose primary field is infrared radiation. Its usage helps place the UV in relation to the IR, but the unit becomes cumbersome in the extreme UV.

Further subdivisions of the ultraviolet are:

Near Ultraviolet—NUV This region extends from the short wavelength limit of human vision to about the short wavelength limit of the solar ultraviolet that reaches the surface of the earth. The limits are approximately 400 to 300 nm.

Middle Ultraviolet—MUV The Mid UV covers the region from 300 to 200 nm, which is approximately the region between the solar short wavelength limit at ground level and the onset of strong molecular oxygen absorption. Most solar radiation in this range is absorbed in the atmosphere by ozone.

Far Ultraviolet—FUV This region extends from about the beginning of strong oxygen absorption to about the limit of availability of rugged window materials, the lithium fluoride transmission limit. The range as used here extends from 200 to 100 nm.

Vacuum Ultraviolet—VUV This region includes wavelengths between about 200 nm and 10 nm. The vacuum in the name refers to the

2.1. THE ULTRAVIOLET DEFINED

fact that ground level instruments are usually placed under vacuum to obtain sufficient light transmission in this region. In this book, FUV and EUV as defined here are preferred.

Extreme Ultraviolet—EUV The extreme ultraviolet, sometimes abbreviated XUV, is defined here as 100 nm to 10 nm. The division between the FUV and the EUV is frequently considered to be the ionization threshold for molecular oxygen at 102.8 nm. The EUV solar radiation is responsible for photoionization at ionospheric altitudes. The division between the EUV and the x-ray regions corresponds very roughly to the relative importance of interactions of the photons with valence shell and inner shell electrons, respectively.

Soft X-ray The soft x-ray region is a term used for the shorter wavelength EUV and longer wavelength x-ray regions. As most frequently used, it is centered between about 10 nm to 1 nm.

The divisions of the ultraviolet described above are in common use, but they are by no means the only divisions that will be encountered. They will be used as much as possible in this book, however. These names are shown schematically in Figure 2.2.

In biology and medicine, the ultraviolet ranges are commonly described as **UV-A** and **UV-B**, with the latter being the range having the most detrimental effects on biological materials. The range of UV-A is from about 400 to 320 nm, and the range of UV-B is from about 320 to 280 nm. The term **UV-C** is used for the region from 280 nm to shorter wavelengths.

The term **DEEP UV** is used in describing applications in microscopy and microlithography. The range of the deep UV is from approximately 350 to 190 nm. The short wavelength end of this range sets the limit for the use of the ultraviolet over the small transmission distances needed in laboratory and industrial applications. This limit is due to strong atmospheric molecular oxygen absorption.

There are two other commonly used terms referring to regions of the ultraviolet that may be confusing. One is **solar-blind**. This term is usually referring to wavelengths in the range 300 to 100 nm. A solar-blind sensor can be used in the ultraviolet without concern about having the sun in or near the field of view. The solar visible is much stronger than the solar ultraviolet and most sources that the sensor is trying to measure. Not only does the sun blind the sensor to other, weaker sources, it is usually so intense that it overloads and destroys the sensor. This usually fatal consequence can

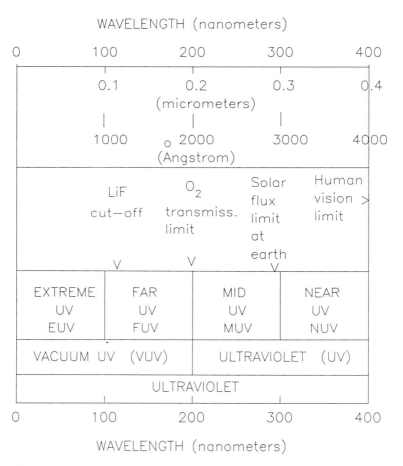

Figure 2.2: Ultraviolet wavelength regions and nomenclature

occur due to the long wavelength tail of the photocathode sensitivity curve. Nevertheless, with special photocathodes as insensitive as possible to the visible and with effective filters, it is possible to have a solar-blind sensor. The solar-blind region defined here covers the MUV and the FUV. In some instances, however, the term is used to mean only the MUV.

Another possibly confusing term is the **windowless ultraviolet**. While thin metallic films may sometimes be used as windows, the availability of rugged, sealed detectors is limited to wavelengths longer than about 105 nm, the lithium fluoride transmission limit. At shorter wavelengths in the EUV, windowless detectors, gas cells, etc., separated by differential pumping systems in vacuum systems or flooded with a gas such as helium, which has little absorption at the wavelengths of interest, have to be used to make measurements. These experimental difficulties have been overcome, and it is possible to make accurate laboratory measurements throughout the ultraviolet without windows. In the low pressures of space, there is usually no problem with using windowless detectors for the EUV.

2.2 Global ultraviolet

When seen from space in ultraviolet wavelengths, the Earth presents a far different picture from the blue sphere with swirling white clouds seen by the first astronauts and now familiar to us as spaceship Earth. At far and extreme ultraviolet wavelengths, each pole of the earth is capped with a glowing "halo" marking the **auroral oval**. The auroral ovals can be seen even in the daytime above the dayglow of the solar-illuminated half of the sphere. Their size and detailed structure change, somewhat like weather clouds, as they follow the sun, responding to changes in the solar wind and to geomagnetic storms.

Other, weaker, emission features can be seen above the planet. On each side of the magnetic equator during early evening hours, bands of FUV and EUV radiation due to the **tropical UV airglow** stretch across the globe. Sensitive imagers can see **polar cap structures**, which are glowing arcs and patches of light that move across the darker polar cap. Even more sensitive instruments can observe weak emissions from the night airglow in many UV wavelengths.

The picture changes depending on the wavelength. The Earth in the hydrogen Lyman alpha line at 121.6 nm is a bright but featureless scattering source that is called the **geocorona**. This type of image results from multiple scattering of the intense solar Lyman alpha emission line. It extends out many earth radii and is of significant intensity also at night, as the radiation scatters around the globe.

In the middle and near ultraviolet, a different sort of picture emerges. When the scattering emission seen from space is displayed as ozone concentrations, seasonal and long-term changes over the globe can be seen. The most dramatic has been the **ozone hole** in the antarctic ozone levels. At other places and times, observations in the MUV and NUV, which can see further down into the atmosphere than the FUV and EUV, reveal volcano plumes and trails from high-flying aircraft.

An image of the earth taken from the moon on the Apollo program is shown in Figure 2.3. The airglow on the day side, the auroral ovals, and the tropical UV airglow belts can be seen in the image, which was taken in the approximately 125 to 170 nm wavelength band of the FUV. This work is discussed further and referenced in Chapter 16.

A schematic view of the approximate locations of global remote sensing regions based on these emission features is given in Figure 2.4. These regions are used to organize the discussion of future chapters.

As a final introductory example of global UV emission, an image of the auroral oval taken by the Polar BEAR satellite is shown in Figure 2.5. This image is obtained from the emission of the oxygen atom at 135.6 nm. The observed image is superimposed on the land mass outlines of the northern polar region. Sun aligned arcs can be seen in the polar cap, along with intricate structure caused by the variability of the incoming energetic particles causing the emission. Auroral imaging is discussed further in Chapter 16.

2.3 Atmospheric ultraviolet

The transmission of ultraviolet radiation in the atmosphere varies from virtually no absorption in the NUV to extremely strong absorption at shorter wavelengths. One way to acquire a general understanding of UV in the atmosphere is through Figure 2.6, which displays the altitude at which solar, or any other, radiation from the zenith direction incident on the atmosphere is reduced to $1/e$ of its original intensity. This is the altitude of unit optical depth (see also Chapter 7). Note that EUV structure has not been included in Figure 2.6.

Figure 2.6 gives the altitudes where the principal deposition of solar radiation occurs in the atmosphere. Most of the absorption of a given wavelength

2.3. ATMOSPHERIC ULTRAVIOLET

Figure 2.3: Ultraviolet image of the Earth as seen from the moon. The FUV wavelength band is from about 125 to 170 nm. (Image from G. R. Carruthers, Naval Research Laboratory, used with permission)

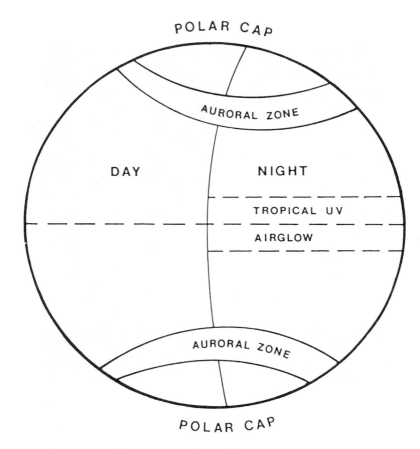

Figure 2.4: General location of UV remote sensing regions

2.3. ATMOSPHERIC ULTRAVIOLET

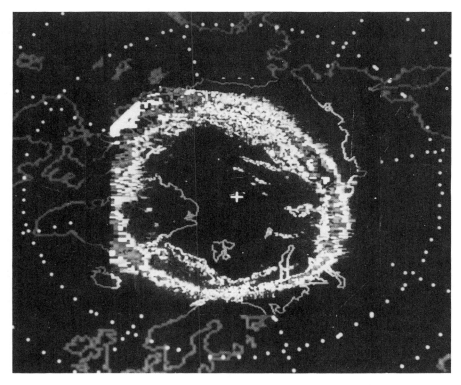

Figure 2.5: UV image of the auroral oval from the Polar BEAR satellite. Emission is from atomic oxygen at 135.6 nm. (R. E. Huffman, F. P. Del-Greco, and R. W. Eastes, Geophysics Directorate, Phillips Laboratory)

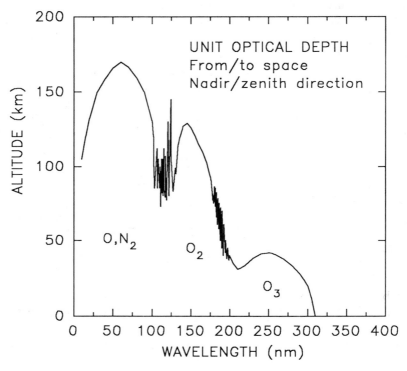

Figure 2.6: Altitude of unit optical depth (1/e absorption) for ultraviolet radiation from or to space in the nadir or zenith direction, respectively, (adapted from Handbook of Geophysics and the Space Environment)

2.4. WHY USE UV?

occurs near the $1/e$ absorption altitude. Also shown are the atmospheric constituents that are most important for the absorption of solar radiation at any wavelength. The absorption processes and the photon cross sections that control this absorption are discussed in detail in Chapters 7 and 8.

In Figure 2.6, it is assumed that the sun is directly overhead, or that the solar zenith angle (SZA) is equal to zero, which is not the case most of the time. When the sun is not directly overhead, the slant transmission path includes more atmospheric absorbers and the $1/e$ absorption altitude is slightly higher. The curve shown is a useful minimum altitude.

Figure 2.6 also provides some idea of how deeply one can see into the atmosphere from space. For an emission source at the altitude given by the curve, the intensity as seen from space looking toward the nadir, or earth center, direction is reduced by $1/e$ of its original intensity by atmospheric absorption. At slightly lower altitudes, or for slant paths, the emission source will be reduced so much by absorption that it cannot be detected. Thus, the curve sets the altitudes where passive remote sensing of the atmosphere is feasible, with NUV and MUV applications in the troposphere and stratosphere and FUV and EUV applications in the ionosphere and thermosphere.

Many of the most important applications of the UV are and will be in space. The absorbing blanket of atmosphere prevents transmission both from and to the surface. However, absorption also uncouples space UV from near surface complications such as scattering from clouds and surface features. Depending on the application, the results of atmospheric photoabsorption may be an asset or a liability.

While Figure 2.6 is extremely useful, it cannot illustrate some features of the ultraviolet and the atmosphere. For example, the primary cause of the solar limit at the surface near 300 nm is absorption by stratospheric ozone. However, for transmission at or near the surface in the 200 to 300 nm region, where there is less ozone in the transmission path and air does not absorb strongly, it is possible to have useful applications over ranges of a few kilometers. In practice, radiance and transmission codes should be used, as discussed in Chapter 13.

2.4 Why use UV?

In many remote sensing applications, there is no choice about the wavelength region or the approach to use for a particular task. The phenomenology determines the approach. There are, however, situations where there is a choice possible among infrared, visible, ultraviolet, and x-ray sensing. In these cases, and for future similar cases that may arise, it is useful to compare and contrast the relative merits of sensing in these wavelength regions. In many applications, it may well be found that the best approach is to use integrated or fused sensors operating in several wavelength regions and providing a more complete picture than any separate sensor.

While the items listed below will be covered in more detail at the appropriate place later, they are discussed together here to compare the advantages and disadvantages of using ultraviolet compared to infrared or other types of sensors. Some of these may not be apparent at first sight.

Advantages of using ultraviolet sensors are as follows:

- Cryogenics are not required. While passive cooling may be used in some UV sensors, the extreme temperatures of liquid helium and liquid hydrogen needed for the most sensitive IR sensors, especially in the LWIR, are not required. The weight and complexity penalties for cryogenics in space are considerable.

- Large focal plane arrays are now available. For example, the second generation of the Hubble Space Telescope will probably have an array of 2000 x 2000 pixels. Large arrays developed initially for the visible region can readily be adapted to the ultraviolet.

- For the same spatial resolution, the smaller wavelengths of the UV allow smaller collecting optics, assuming that the emission from the source is equivalent. The collection optics can have a diameter smaller by about a factor of ten because the UV wavelengths are smaller than SWIR wavelengths by about this factor. This decrease in the diameter leads to decreases in the weight and volume of a space sensor by a total factor of about 500 times.

- Because of the previously mentioned advantages, UV sensors are usually smaller and weigh less than equivalent IR sensors. They are thus generally less expensive to launch and use in space.

Disadvantages with the use of UV, at least at the present time, are:

- Radiance and transmission models are not as well developed as in the visible and IR, although the situation is improving. The extension of the visible/IR code LOWTRAN 7 through the MUV is an example of a recent improvement.

2.4. WHY USE UV?

- The nature and magnitude of any clutter contribution to the UV background is almost totally unknown. New measurements are needed.

- Uses are generally more dependent on atmospheric transmission than in the visible and IR. This sometimes can be converted into an advantage, but it does remain as a complication that must be taken into consideration.

Perhaps the most important point in the use of the ultraviolet for applications in space is that the weight and size of the sensor can be made relatively small. The satellites or other space platforms can thus be made lighter in weight and therefore less expensive. This fact makes possible constellations of sensors in space at a cost that is within consideration. Until there is some fundamental reason why ultraviolet sensors cannot be used, they should be regarded as a candidate. The disadvantages, especially relating to further measurements and models, are steadily being overcome.

Chapter 3

Radiometry

In order to explain the use of ultraviolet radiation for remote sensing, it is necessary to introduce a few radiometric and optical principles. The emphasis will be on measurements and sensors. Therefore, the equations will be developed with the idea that they will be used for initial design and feasibility considerations and for understanding the physics of remote sensing applications.

3.1 Photons, watts, and Rayleighs

The energy possessed by a photon of electromagnetic radiation is given by Planck's relation

$$E_\lambda = h\nu' = \frac{hc}{\lambda} = hc\nu. \tag{3.1}$$

In the equation, E_λ is the energy of the photon, or quantum, of electromagnetic radiation, h is Planck's constant ($6.6260755 \times 10^{-34}\ joule\ sec$), ν' is the frequency of the photon in sec^{-1}, c is the speed of light ($2.99792458 \times 10^{10}\ cm/sec$, λ is the wavelength in cm, and ν is the wavenumber in cm^{-1}. From this equation, it can be seen that the wavenumber can be used as an energy unit, as is commonly done in spectroscopy. The equivalent wavenumber for a wavelength of 100 nm, or 1000 Å, is 100,000 cm^{-1}.

The above equation gives the energy of a photon in joules, but it is more convenient for us to use the electron Volt. The conversion factor is as follows:

$$E_\lambda = 1.6022 \times 10^{-19} E_J. \tag{3.2}$$

In this equation, E_λ is in electron Volts and E_J is in joules. It is frequently necessary to know the energy equivalent of a photon of a specific wavelength.

This energy is given by the following equation:

$$E_\lambda = \frac{12399}{\lambda}. \qquad (3.3)$$

In this equation, E_λ is again in electron Volts and λ is in Å. Thus, a photon with a wavelength of 100 nanometers or 1000 Å has an energy of about 12.4 eV.

Energy units for the range covered by the ultraviolet are shown in Figure 2.1. The range 400 to 10 nm covers approximately 3 to 120 eV. Photoabsorption and photoemission processes involving atmospheric atoms and molecules at these energies are associated with excitation to upper states, photodissociation, and photoionization.

The radiant flux in a given wavelength interval $\Delta\lambda$ can be given in watts, which is joules per second, or in photons per second. A useful relation between these two is as follows:

$$W = \frac{1.98645 \times 10^{-15}}{\lambda} \Phi \qquad (3.4)$$

In this equation, W is in watts, λ is in Ångstroms, and Φ is in photons per second. A useful number to remember is that one watt is about equal to about 10^{18} photons per second at 2000 Å.

The **Rayleigh** has been adopted as a photometric unit in airglow and auroral measurements. It was originally proposed as a measure of the emission from a column of atmosphere along the line of sight of a measurement, but it is frequently generalized to be an apparent radiance unit. The quantity of primary interest in the study of airglow and aurora is the volume emission rate over the altitude range of interest. However, multiple scattering, absorption, and possibly other complications may prevent a simple relationship between the column emission rate as measured and reported in Rayleighs and the volume emission rate.

The Rayleigh is defined as

$$\text{One Rayleigh} = \frac{10^6 \text{ photons}}{\text{sec cm}^2 \, (4\pi \text{ sterad}) \, (\Delta\lambda)} \qquad (3.5)$$

The form of the definition given here is for its use as a generalized unit of apparent radiance, as given by *Baker*, 1974 and *Baker and Romick*, 1976, who discuss the historical and the generalized definitions of the Rayleigh unit.

3.2. RADIANCE AND IRRADIANCE

In the historical definition, the cm^2 in the denominator is meant to be interpreted as cm^2 (column), but as mentioned above it may be used like the unit area of an emitting surface in a radiance, B. In the historical definition, the 4π steradian unit in the denominator of Eq. 3.5 is omitted. In practice, the Rayleigh may be used simply because it is a convenient way to describe the result of a measurement. Without additional experimentation or detailed modeling, undisturbed column emission along the line of sight cannot be assumed.

The $\Delta\lambda$ in the denominator indicates that the value given is over a wavelength interval. For a prominent airglow or auroral feature, the wavelength range unit is usually left out, and the value given in Rayleighs is for this "feature" in the spectrum. For continuum sources, the use of a wavelength interval gives the emission level. The wavelength interval unit used may be for convenience or for comparison with other measurements. It also has no relationship to the resolution of the measurement, which is usually set by a spectrometer or a photometer filter. The kiloRayleigh is frequently employed, especially in auroral investigations.

The Rayleigh unit was introduced by *Hunten et al.*, 1956, and named for the fourth Lord Rayleigh, who did pioneering airglow and auroral research. Rayleigh scattering, also of great importance for ultraviolet remote sensing, is named for his father, the third Lord Rayleigh. An excellent discussion of the rationale for the introduction of the Rayleigh is found in *Chamberlain*, 1961. It is the generally accepted unit for airglow and auroral intensities in the visible and ultraviolet.

3.2 Radiance and irradiance

The precise consideration of radiation emitted from surfaces and point sources and then incident on apertures or other surfaces has led to a proliferation of names, units, and symbols. An excellent recent discussion with references is given by *Spiro and Schlessinger*, 1989. The simplified nomenclature used here is based primarily on the preference of the author. The interested reader is also referred to *Wyatt*, 1978, 1987; *Grum and Becherer*, 1979; and *Boyd*, 1983, for more detailed discussions of radiometry.

Nomenclature used for the ultraviolet can be simpler than the infrared because equilibrium radiation is less important in the ultraviolet. Only the terms to be used in further discussions in this book will be introduced here. No use will be made of the luminous flux approach, and the discussion will concentrate on photons, or quanta, per second and watts. Also, extended sources will be considered to be Lambertian, or totally diffuse. As is the case with all language, usage eventually will determine nomenclature, and the pronouncements of authorities will fade away if not adopted in practice by a community of users.

The **radiant intensity** (J) is the radiation emitted per unit solid angle and over a given wavelength interval by a point source or by a source at a sufficient distance from the observer so that it can be regarded as a point source. An example is the sun, which for atmospheric use is regarded as a point source with the full-disk emission incident on the Earth's atmosphere of primary interest. The units of radiant intensity are either $watt/steradian\,(\Delta\lambda)$ or $photon/sec\,(steradian)\,(\Delta\lambda)$. For a given sensor, a source may be regarded as a point source if the image of it on the focal plane is small relative to the size of the sensitive area of a detector or to the pixel (picture element) size in an array. Another name for radiant intensity is pointance.

The **irradiance** (F) is the radiation per unit area and given wavelength interval incident on the entrance aperture of a sensor or incident at some other designated place. For application of solar flux measurements to atmospheric problems, for example, tables most often use irradiance rather than the radiant intensity or the radiance of the sun itself. The units of irradiance are $watt/cm^2\,(\Delta\lambda)$ or $photon/sec\,cm^2\,(\Delta\lambda)$. The irradiance is sometimes called the areance.

The **radiance** (B) is the radiation in a given wavelength interval from an extended source per unit area and unit solid angle. The B might be considered to stand for background in this book, although it can in principle refer to any extended source. The airglow can be considered an extended source, characterized by a radiance, when seen from a distance. The units are $watt/cm^2\,steradian\,(\Delta\lambda)$; $photon/sec\,cm^2\,steradian\,(\Delta\lambda)$; or $Rayleigh/(\Delta\lambda)$. Radiance is sometimes called the surface brightness or the sterance.

When dealing with radiation, a wavelength band or interval is always involved, although for various reasons it may not be explicitly given. The units of J, B, and F therefore have a $\Delta\lambda$ in the denominator. In practice, where the wavelength interval is considered to be understood, it is sometimes omitted in the units. The units are sometimes given as a level of radiation over an arbitrary larger range for convenient comparisons. A common unit for radiance is $watt/cm^2\,sterad\,micrometer$, although the micrometer unit is really too large to use as an ultraviolet wavelength interval. Sometimes

3.3. RADIOMETRIC EQUATIONS

the wavelength interval is called the wavelength band, the bandwidth, or just the band.

The relationship of units commonly used in radiometry can be found with the equations given in this chapter. For many engineering and infrared applications, radiance units using $\mu watts$ are preferred. For airglow and auroral investigations, Rayleigh units are common. In order to facilitate conversion between the two types of units, the following equation is given:

$$B_{\mu w} = \left(\frac{1.5808}{\lambda}\right) B_R. \qquad (3.6)$$

$B_{\mu w}$ is the radiance in units of $\mu watt/(cm^2)(steradian)(\mu meter)$; λ is the wavelength in Ångstroms; and B_R is the radiance in units of Rayleighs/ Ångstrom. As an example, both of these units are used as units for atmospheric backgrounds in Figure 12.1.

3.3 Radiometric equations

The consideration of remote sensing methods requires that the entity being sensed provide a signal from a sensor located some distance away. It is best to initially determine the response of the sensor in counts or similar units before considering the signal-to-noise, signal-to-background, and other aspects of the measurement. This approach will keep attention focused on the key issue of the number of signal counts. Therefore, the following discussion considers the counts per second from a sensor for different types of sources. The sampling period will then determine the number of counts and associated statistical uncertainty, as discussed in Section 3.5.

The equations given here are only those useful for our purposes and are given without the proof found in references on radiometry.

Ultraviolet sensors usually give outputs in counts, count equivalents, or currents. Many ultraviolet detectors utilize some form of the photoelectric effect, and single photon counting is commonplace. Note that not all photons entering the aperture are counted but that each count is produced by a single photon which has survived reflections, dispersion, photocathode efficiencies, etc., before being recorded as a count.

Radiant intensity, J, from a point source will produce the following output in counts per second from a sensor:

$$\frac{C}{t} = J(\Delta\lambda)\left(\frac{A}{R^2}\right)EK. \qquad (3.7)$$

Table 3.1: K values for radiometric equations

Radiometric units	K (λ in Å)
photons	1
watts	$(5.034 \times 10^{14})\lambda$
Rayleighs	$10^6/4\pi = 7.958 \times 10^4$

In this equation, C is the number of counts; t is the counting period in seconds or other time units; J is the radiant intensity in units such as $photon/(sec)(sterad)(\Delta\lambda)$ or $watt/(ster)(\Delta\lambda)$; $\Delta\lambda$ is the wavelength interval in any wavelength unit such as nanometers, Ångstroms, or micrometers; A is the effective area of the collecting aperture or collector; R is the range or distance from source to sensor; E is the total efficiency of the sensor in counts out per photons in; and K is a constant which varies with the units (photon, watt, Rayleigh, etc.) used for J.

Table 3.1 provides values of K to be used for the radiometric equations in this chapter.

In Eq. 3.7, the ratio of the aperture to the slant range squared is shown in parentheses to emphasize that this quantity is the solid angle in steradians over which radiation is collected from the source, which is assumed to be emitting isotropically over the total 4π steradians of solid angle.

The total efficiency, E, is a fraction which characterizes the sensor as a radiation measuring device. It is the radiometric calibration of the total sensor. It includes all the reflections, quantum efficiencies, and other interactions of the photons with the sensor that occur between input and output. These factors, which are crucial to the success of a remote sensing method, will be described in more detail when specific sensors and techniques are discussed in Chapter 4. At present, we can use a range of 0.001 to 0.1 to cover sensors of interest here, with 0.01 a rough middle value.

While the total efficiency, E, may be approximated from the design of the sensor and the nominal specifications of the components, it should always be determined for the sensor as a whole by some kind of calibration before results are reported. There are other factors that are assumed to be satisfied. For example, the use of E in the equations assumes that the detector is in a linear region of its dynamic range and well away from saturation. Otherwise, a more complicated dependence of the output on the input will be needed in equations.

It is also assumed that the wavelength band is well isolated so that there

3.3. RADIOMETRIC EQUATIONS

is no contribution from out-of-band emission. This "red leak" problem is not trivial for many UV measurements, and lack of attention to this possible error has compromised a number of measurements. It is discussed further in Chapter 4.

Irradiance, F, is used at times as a more convenient way to describe the flux incident upon a sensor. The output of a sensor in this case can be found from the point source equation. For a point source, the radiant intensity is related to the irradiance F as follows: $J = FR^2$. Substituting this into the equation for radiant intensity given above gives the equation for use with irradiance units.

$$\frac{C}{t} = F(\Delta\lambda)AEK. \qquad (3.8)$$

The irradiance, F, can be in $photons/(sec)(cm^2)(\Delta\lambda)$ or $watts/(cm^2)(\Delta\lambda)$. The remaining symbols and units are as given previously in this chapter. This form of the equation is useful for solar flux measurements, as solar radiation is usually given in irradiance units (see Chapter 7).

Radiance, B, from an extended or diffuse source larger than the field of view of the sensor detector or pixel will give the following output from a sensor:

$$\frac{C}{t} = B(\Delta\lambda)A\Omega EK. \qquad (3.9)$$

The radiance, B, can be in units such as $photon/(sec)(cm^2)(ster)(\Delta\lambda)$; $watt/(cm^2)(ster)(\Delta\lambda)$; or $Rayleigh/(\Delta\lambda)$. Except for Ω, which is in steradians, the symbols in the equation have been previously defined.

The field of view (FOV), Ω, in steradians, is the solid angle observed by the sensor, and it is determined by the design and fabrication of the sensor. The field of view Ω is sometimes called the object field of view. The area observed, S, from an extended source at a slant range R is ΩR^2. We assume in the equation that the field of view Ω is well-defined for the sensor, with good rejection of radiation not in the field of view, but this should be proven by calibration and test.

The above equation for an extended source using the object FOV and the sensor aperture has been found to be the most useful formulation when considering applications. It emphasizes the size of the entrance aperture, which frequently determines the weight and size of the total sensor package.

There is an alternate approach using the *f-number* and pixel area of the sensor that is also frequently used. This form of the radiance equation can be developed by using the general principle that the *throughput* of an optical system is invariant. Thus, the following equality is valid:

$$\text{(collector area)(object field)} = \text{(pixel area)(pixel field)}$$

The left side, outside the sensor, is the product of the effective area of the sensor aperture, A, and the object field of view, Ω, in steradians. The right side, inside the sensor, is the area of the sensor pixel, a, and the pixel field of view, ω, in the steradians that this pixel covers at the focal plane of the sensor. The throughput equality can be given in symbols as follows:

$$A\Omega = a\omega. \tag{3.10}$$

The throughput is also called the ètendue.

The *pixel*, which is an abbreviation for picture element, is a term borrowed from image analysis as a convenient way to refer to a single member of an array detector. For a detector with a single active area, such as a photomultiplier tube with a single photocathode, the pixel could be considered to be the total photocathode and the pixel area is its area. For a one or two-dimensional array, a pixel is an individual element in the array and the pixel area is the sensitive area of this individual element. For currently available charge coupled devices (CCD) and other types of arrays, the pixel usually has linear dimensions measured in micrometers, and there may be thousands of pixels in the array.

The *f-number*, frequently employed in optical sensor design, is defined as follows:

$$\textit{f-number} = \frac{f}{D}, \tag{3.11}$$

where f is the focal length of the sensor and D is the diameter of the collecting optics. The pixel field of view ω is equal to the following:

$$\omega = \frac{A}{f^2} \tag{3.12}$$

$$= \frac{\pi D^2}{4f^2}. \tag{3.13}$$

Finally, the **radiance** equation using the *f-number* is given by rearranging the above and substituting it in Eq. 3.9 to yield:

$$\frac{C}{t} = B(\Delta\lambda)\left(\frac{\pi a}{4(\textit{f-number})^2}\right) EK, \tag{3.14}$$

which shows that the counts per second are proportional to the inverse square of the *f-number*. Both forms of the radiance equation have uses in consideration of remote sensing concepts.

The radiometry equations given here assume idealized sensors, Lambertian emission from extended sources, and circular optics. The equations may need to be adapted for specific situations. If an object being measured has an image in the focal plane large compared to the individual pixel size, it can be considered an extended source for that pixel, and the formulation of the equation with the *f-number* will be found to be useful.

The use of the equations given in this section to define the sensitivity of remote sensing instruments is considered in Chapter 4. Typical values found in practice will be given there.

3.4 Spatial considerations

There are two simple but important concepts from basic optics that should be kept in mind in regard to remote sensing, especially from space. These are the relationship of the spatial resolution of the sensor to its focal length and the relationship of the diffraction limit, which is roughly the best angular resolution of the sensor, to the size of the aperture. The term spatial resolution is used here loosely to mean the length dimension of the object as determined by geometric optics.

The relation between the size of the object and the size of the image from elementary geometrical optics is as follows:

$$\frac{O}{R} = \frac{i}{f}, \qquad (3.15)$$

where O is the length of the object that is observed; R is the slant range from the object to the sensor; i is the pixel linear dimension; and f is the sensor focal length. This equation sets the size of the focal length, which needs to be minimized for all space and many ground-based sensors.

As an illustrative example, the typical pixel size of ultraviolet and visible focal plane arrays based on charge coupled detectors (CCD) or similar devices is currently about 20 micrometers along each side of a square pixel. Observation of a one meter object at a range of 1000 kilometers would require a 20 meter focal length. This focal length is possible on the shuttle and large space structures such as a space station, but not on small satellites. Through the use of folded optics, large focal lengths can sometimes be achieved within compact overall dimensions but at the expense of total sensitivity. It is obvious that this kind of resolution is difficult to achieve. The spatial resolution most suited to small satellite sensors is about 100 meters from the same 1000 kilometer range, which gives an achievable 20 centimeter focal length.

The diffraction limit of the optics employed must also be considered. A convenient measure of the limiting angular resolution of the sensor is based on the location of the first minimum in the diffraction pattern of a point source at the focal plane of an optical instrument. The circular pattern defined by this minimum is called the Airy disk. The angular resolution α in radians is given as follows:

$$\alpha = 1.22 \left(\frac{\lambda}{D}\right). \tag{3.16}$$

In this equation, λ is the wavelength of light utilized and D is the diameter of the collecting aperture in the same units.

The spatial length of the object is the angular size in radians times the slant range. For the case of a one meter object at a range of 1000 kilometers used previously, an angular size of one microradian, or 10^{-6} radians, is needed. If the image is obtained in ultraviolet radiation with a wavelength of 200 nanometers, or 2000 Å, the aperture diameter required is about 24 centimeters, which is easily achievable. To obtain the same angular resolution in the infrared at a wavelength of 2 micrometers, the aperture grows by a factor of ten to 2.4 meters, or almost 100 inches, which is about the size of the primary mirror on the Hubble Space Telescope as launched by the shuttle. While such a large collector is therefore obviously possible, it is not simple or cheap.

It is clear that the smaller size of the ultraviolet wavelengths is an advantage in obtaining the highest spatial resolutions, all other things being equal. As is usually the case, other factors intervene, including possibly a relatively low radiance for the ultraviolet source, and so an iteration over all characteristics of the source must be done before a useful sensor can be designed.

3.5 Photon statistics

In the ultraviolet, measurements are usually made in counts or count equivalents, and accuracy is dependent on obtaining sufficient signal counts. In addition, the signal counts must be distinguished from counts due to background UV emission, sensor dark count, and various miscellaneous noise sources. The arrival of counts at the sensor counter is random in nature,

3.5. PHOTON STATISTICS

and Poisson statistics apply. For numbers of counts of interest to us, the fractional uncertainty in the number of counts found in a given sample period is given by the well-known equation:

$$\text{fractional uncertainty} = \frac{\sqrt{C}}{C}. \tag{3.17}$$

The product of the fractional uncertainty times the number of counts is $\sqrt{C}$, which is the standard deviation in a group of samples all collected over the same counting period. Thus, for total counts of 100, the fractional uncertainty expressed as a percentage is 10 %, and to achieve 5 % statistical uncertainty it is necessary to have 400 counts.

The total number of counts received is the sum of signal and unwanted counts, as follows:

$$C_T = C_S + C_B + C_D + C_M. \tag{3.18}$$

In this equation, C_T is the total number of counts in the sample, and C_S is the number of counts due to the signal, or the object of the measurement. The rest of the terms on the right side of the equation are the unwanted counts. The background counts, C_B, may be due to natural background from atmospheric scattering, airglow, aurora, etc., detected along with the signal. The dark count, C_D, is the count from the sensor when not illuminated and under benign environmental conditions—for example, in a laboratory setting. Finally, C_M is made up of all the miscellaneous additional counts that may be encountered under the environmental conditions of the measurement—for example, in space. It may be possible to ignore some of these contributions to the total count in any specific application or through the design of the measurement.

The background count, C_B, is usually from radiation coming from the atmosphere. It is possible that at least a part of this emission is the signal for some applications and the background for others. For example, auroral radiation may be the desired signal, and airglow radiation would be the undesired background for an auroral detection technique.

The dark count, C_D, depends primarily on the type of detector used. The photocathodes used over much of the ultraviolet have extremely low intrinsic thermionic emission, which leads to low dark counts at room temperature, and so the dark counts may be insignificant. The dark count may be decreased greatly by cooling the sensor, if necessary.

Both dark count, or current, and the background counts may be sufficiently large to require some type of phase sensitive detection or "light chopper". In this way, the signal can be seen above a large, steady level due to background and dark current contributions. Although it can be employed if needed, the use of phase sensitive detection is not common in the ultraviolet.

Miscellaneous counts, C_M, may come from the space radiation environment encountered by a satellite, such as in the South Atlantic anomaly or in the particle precipitation regions that form the auroral zone; from light scattered into the detector from a bright out-of-field source such as the sun or an antenna or other spacecraft structure that scatters sunlight into the sensor; from on board radio frequency interference (RFI), which is due to voltage transients from other spacecraft equipment that are coupled into the detector electronics and appear as counts; or possibly from other causes. These counts can usually be decreased by improvements in the baffling and shielding of the sensor and in the design and electromagnetic isolation of the various parts of the spacecraft.

The nonsignal counts should be measured independently and subtracted from the total counts to get the signal counts. One way to determine some of these counts is to have a shutter at the entrance aperture or entrance slit of a spectrometer which can be closed to prevent radiation entering the sensor. The counts measured with the shutter closed are the dark counts, C_D, plus the miscellaneous counts, C_M, not due to light. To be as valid as possible, the determination of nonsignal counts should be made under conditions as similar to the measurement as possible. For example, laboratory dark counts should not be assumed to be the same as the nonsignal counts for a satellite sensor.

It should be kept in mind that a sampling period is needed to determine the number of counts from the count rate given by the equations in Section 3.3. For space applications, satellites are usually moving relative to ground location, unless special pointing is available to "stare" at a scene and thus build up the number of counts. Thus, the number of counts may be largely determined by the amount of time the signal source is in the FOV or by the desired spatial resolution of the measurement.

The **signal-to-noise ratio**, or S/N, is used widely as a measure of detectability, assuming that there are sufficient signal counts to meet the statistical uncertainty requirement. This is the ratio of the signal to the total noise, which is the uncertainty in the total counts, or the square root of the total counts. Using the above equations, the signal to noise ratio,

S/N, is as follows:

$$\frac{S}{N} = \frac{C_S}{\sqrt{C_S + C_B + C_D + C_M}}. \tag{3.19}$$

Note that this equation becomes equivalent to the statistical uncertainty equation when the dark, background, and miscellaneous counts are insignificant compared to the signal counts.

The **signal-to-background ratio**, or S/B, is the important special case when the background dominates the noise. While the dark and miscellaneous counts can be reduced by hardware designs, there is usually no alternative to observation of the source together with the natural background. Then, the question becomes the ratio of the signal to the noise due to the background.

It should be obvious that if there is a constant background, a signal level smaller than the background level can be measured. What is important in detection of a signal is the ratio of the signal to the noise, in this case the square root of the background counts. Many clever methods have been developed to extract signal buried in noise, and thus the ratios S/N and S/B cannot be used to absolutely rule out measurements. They can however give some idea of how difficult it will be.

The effects of the background can be decreased by array detectors with pixel sizes similar to or smaller than the desired spatial resolution. The background to be considered relative to the signal is thus reduced to the angular extent of the pixel. The background is still of considerable importance, however, since detection usually depends on a comparison with the other pixels in the field of view, which will have counts due to the natural background. Also, the background emission is still being detected across the total FOV of the sensor, which may lead to saturations and nonlinearities for some sensors.

Clutter, or geophysical variability, may in practice be a noise source. Little is known about it in the ultraviolet, as discussed in Chapter 12.

3.6 References

Baker, D. J., Rayleigh, the unit for light radiance, *Appl. Optics, 13*, 2160–2163, 1974.

Baker, D. J. and G. J. Romick, The Rayleigh: interpretation of the unit in terms of column emission rate or apparent radiance expressed in SI units, *Appl. Optics, 15*, 1966–1968, 1976.

Boyd, R. W., *Radiometry and the Detection of Optical Radiation*, Wiley, 1983.

Chamberlain, J. W., *Physics of the Aurora and Airglow*, Academic Press, 1961.

Grum, F. C. and R. J. Becherer, *Optical Radiation Measurements I. Radiometry*, Academic Press, 1979.

Hunten, D. M., F. E. Roach, and J. W. Chamberlain, A photometric unit for the airglow and aurora, *J. Atm. Terr. Phys., 8*, 345–346, 1956.

Spiro, I. J. and M. Schlessinger, *Infrared Technology Fundamentals*, Dekker, 1989.

Wyatt, C. L., *Radiometric Calibration: Theory and Methods*, Academic Press, 1978.

Wyatt, C. L., *Radiometric Systems Design*, MacMillan, 1987.

Chapter 4

Sensors

This chapter introduces some convenient measures of sensor sensitivity and describes a representative photometer, spectrometer, and imager used for atmospheric remote sensing. These sensors are used here as examples that illustrate techniques. Calibration both in the laboratory and in space is also introduced.

General references to techniques for the UV have been given by *Samson*, 1967; *Hennes and Dunkelman*, 1966; and *Zaidel' and Shreider*, 1967. Progress since this time has been rapid, with the introduction of electronic imaging techniques and the availability of synchrotron radiation sources. Chapter 1 gives references to recent SPIE conferences on ultraviolet technology and systems and also to the most recent Conference on Vacuum Ultraviolet Radiation Physics. These sources should be consulted for some insight into the current state of development.

This discussion is meant to be illustrative, with the intent of demonstrating principles needed in the understanding of any sensor. Methods for the design of optical sensors are not given. For this purpose, there are a number of recommended references including *Smith*, 1990, *Williams and Becklund*, 1972, and *Malacara*, 1988.

What can be accomplished with sensors at any time is related to the current state of development of UV technology. The UV experimenter should stay in touch with vendors in the UV technology field, including manufacturers of detectors, filters, gratings, and windows, as well as producers of complete UV sensors. Literature from these sources, which is readily available, is an important source of information.

4.1 Sensitivity

It is useful to have common definitions of sensitivities of sensors making radiometric measurements within desired spatial, spectral, and temporal objectives. The approaches given here have been found to be useful in practice and will be used throughout this book.

Equation 3.9 gives the counts per second from a sensor observing an extended source. It can be rearranged to define the sensitivity, S_B, as follows:

$$S_B = \frac{C}{B(\Delta\lambda)t} \quad (4.1)$$
$$= A\Omega E K. \quad (4.2)$$

where the symbols have the definitions given in Chapter 3. Units for S_B can be counts/Rayleigh-second or similar units based on watts or on photons per second. As seen from the equation, S_B depends on the properties of the sensor including the aperture, A; the field of view (FOV) observed, Ω; and the efficiency of the sensor in converting input photons to counts or count equivalents, E. Since E is a function of wavelength, the sensitivity S_B is also a function of wavelength.

The definition of the sensitivity in counts/Rayleigh-second emphasizes the number of counts per second for a given source in Rayleighs. Thus, the number of seconds required to obtain a sufficient number of counts to satisfy the precision desired can be obtained. Also, consideration of the counts means that it is more readily apparent whether the sensor is within its useful dynamic range and not either saturated or providing almost no counts.

The use of counts/Rayleigh-second is preferable to giving sensitivity information in Rayleigh-seconds, which usually means Rayleigh-seconds per count. It is easy to overlook the fact that the value is for only one count, which leads to an overly optimistic idea about the capability of the sensor to make a meaningful measurement, as most measurements obviously require more than one count.

Another unit similar to S_B can be defined for a point source, by a similar rearrangement of Eq. 3.7. In this case, the comparisons must be done with the same slant range. Sensor sensitivities should be carefully defined and complete. The number of photons per second into the sensor (within the aperture and the solid angle of the sensor) that yields one count is sometimes given. This number should be compared to the efficiency, E, as defined here, and it can be converted to S_B as defined here, provided A and Ω are known.

For arrays, the definitions must specify if they are for the total or for the individual pixel field of view.

Discussions of sensitivity sometimes refer to a given *signal-to-noise ratio*. In this case, the assumption is that the only noise is the noise in the signal, and thus to specify the signal-to-noise ratio specifies the number of counts. For example, a signal-to-noise ratio of 10 would mean total counts of 100 and a standard deviation of 10%, using the definitions given in the last chapter.

Various other measures of sensor sensitivity can be defined. The noise equivalent flux density, or NEFD, and the noise equivalent radiance, or NER, are sometimes useful. These are widely used in the infrared, but they are less useful in the ultraviolet, because of the generally smaller sensor noise levels encountered.

Spatial, spectral, and temporal requirements must all be considered in the choice of a sensor. Sometimes these must be modified to obtain the performance desired. For example, relaxing the spatial and spectral requirements will enable the temporal need for a given measurement to be relaxed. As space platforms are usually in motion compared to the atmosphere, the temporal need may easily become the dominant factor. This is usually not the case for ground-based and laboratory sensors.

It is especially important for sensors used in satellites to be as compact as possible. Thus, an understanding of the prioritized needs of the measurement task and the contributions of each part of the sensor to the sensitivity must to be known. The diameter of the entrance aperture is crucial because as it increases, the other two dimensions of the sensor also increase. A useful rule of thumb is that the size and weight of a space sensor is roughly proportional to 500 times the aperture diameter, rather than 1000 times if the volume increase is exactly scaled with the aperture diameter increase.

4.2 Photometer

A photometer, or radiometer, measures the radiance of an extended source or the radiant intensity of a point source over a wavelength band for the total field of view (FOV) of the sensor. The wavelength band is usually isolated with interference or other types of transmission filters and the wavelength response of the photomultiplier or other type of detector.

A photometer flown on the S3-4 satellite to measure VUV backgrounds, *Huffman et al.*, 1979, 1980, is shown in Figure 4.1. It has an EMR 542G-09

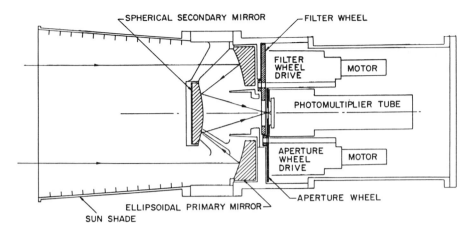

Figure 4.1: Photometer used on S3-4 satellite, *Huffman et al.*, 1980

photomultiplier as the detector operating in the photon counting mode, with a counting period of 10 milliseconds. The photocathode is cesium iodide, and the window is magnesium fluoride. The MgF_2 window limits the lower wavelength sensitivity to about 120 nm, and the CsI photocathode is useful from this lower limit to about 190 nm in this case.

A *photomultiplier* tube converts the photons arriving at its photocathode to electrons, either in the form of a current, for larger flux levels, or in the form of counts, as almost exclusively used for remote sensing flux levels. The tubes used in the S3-4 flight had the photocathode deposited as a thin, semitransparent film on the inside of the window material in a sealed tube, which is called a semitransparent photocathode. Other arrangements have the photon beam striking an opaque thickness of the photocathode material.

Common photocathode materials for the UV include cesium iodide, cesium telluride, potassium bromide, and trialkali, to mention some of the most common. The first three are called "solar-blind" photocathodes. They have peak response in the FUV or MUV and then drop rapidly in sensitivity toward longer wavelengths. For many problems where sunlight is present or possibly present, the solar-blind response is not adequate to prevent a high flux of photons in the visible region to cause an unacceptable "red leak" or out-of-band signal. Special filters or techniques may have to be employed to overcome this problem.

4.2. PHOTOMETER

In the EUV, where there are not rugged windows, photocathodes of metals or secondary emission surfaces, such as the walls of microchannel plates, are used.

After photoelectrons are released from the photomultiplier cathode, they are amplified by secondary emission in a string of dynodes inside the tube. The primary photoelectrons are accelerated to several hundred volts and then collide with special surfaces that have high secondary emission, so that each electron striking the surface releases several electrons on the average. These are accelerated to additional stages, which might finally total fifteen. The total voltage across the dynode chain may be several thousand volts, and the total gain is typically in the 10^5 to 10^7 range. Either counts or a current can be obtained from the tube, with count rates up to several megahertz possible. There are number of types of photomultipliers available.

Other similar detectors may be encountered. The UV *photodiode*, made with similar windows and photocathodes but without the dynodes for secondary emission gain, finds use as a secondary standard detector. The channeltron is a compact photomultiplier with only one channel. It can incorporate any of the UV photocathodes or be used in the EUV with no photocathode.

The filter wheel of the S3-4 photometer has five positions with interference filters centered at 121.6, 134.0, 155.0, and 175.0 nm and a no filter, or open, position. The aperture wheel also has five positions, one closed and four with circular apertures giving angular fields of view from a maximum of $6.9°$ to a minimum of $0.12°$.

The two-element f/1 telescope consists of an ellipsoidal primary collecting mirror and a convex spherical secondary mirror (Dall-Kirkham system) which focuses the field through a nonlimiting hole in the collector mirror onto the chosen aperture. The effective collecting area is about 300 cm^2, corresponding to a clear aperture diameter of about 20 cm. The optics surfaces are evaporated aluminum with a magnesium fluoride overcoating, which is widely used in the far ultraviolet. A sunshade helps prevent unwanted light from entering the sensor.

The sensor was calibrated and tested both at the component level and as a complete sensor. The calibration as a complete sensor is especially important to assure that the filters and apertures are as desired. The transmission peak and shape of interference filters change as the incident angle of the radiation changes, and for divergent or convergent rays, the best approach is a calibration as it will be flown.

A photometer of this type can be very sensitive when observing an extended source such as the airglow, and a wide dynamic range can be achieved using the aperture wheel. For the largest aperture of $5.9°$, the sensitivity, S_B, in counts/Rayleigh-second is as follows: 121.6 nm: 714; 134.0 nm: 1220; 155.0 nm: 208; 175.0 nm: 143; and no filter: 6250. Thus, radiance values of less than a Rayleigh measured in less than a second are possible. The data stream is recorded at each 10 millisecond counting period, and then counts may be summed over intervals of interest.

The laboratory dark count for this sensor was 0.5 counts per second. In flight, the nonauroral dark count was 4 counts per second, while the dark count in the auroral zone was 25 counts per second. These dark counts were usually insignificant compared to the airglow and auroral counts.

The photometer described here is for the far ultraviolet region, but with changes in filters and photomultipliers, it could be used in the middle ultraviolet or near ultraviolet as well. The extreme ultraviolet, at wavelengths shorter than about 100 nm, is more difficult, as there are not rugged window materials below the LiF transmission limit at about 105 nm. Reflective coatings needed are also different, but they may be efficient enough to use normal incidence methods down to about 50 nm. Below this point, the use of grazing incidence may be necessary. Recently, the use of multilayer coatings has allowed consideration of the much simpler normal incidence approaches at wavelengths up to about 30 nm, in the lower wavelength part of the EUV range.

The photometer in Figure 4.1 is fairly elaborate. Frequently, simple photometers with a single fixed filter are used to provide a measure of an important emission feature or to provide redundancy in a flight instrumentation package.

For use in Eq. 4.1, A is the clear aperture of the mirror, and Ω is determined by the size of the detector and the focal length. The efficiency E is obtained as described in Section 4.5.

4.3 Spectrometer

A spectometer or spectrograph provides measurements over a range of wavelengths. The wavelength band isolated is generally smaller than for a filter photometer. The common element in spectrometers and spectrographs is some method to disperse the entering radiation as a function of wavelength. In the ultraviolet, this common element is almost always a diffraction grat-

4.3. SPECTROMETER

ing, but interferometers are possible in the NUV and MUV.

Current usage refers to a **spectrometer** as a sensor with a single detector such as a photomultiplier, with mechanical scanning of the spectrum across an exit slit placed before the detector. A **spectrograph** obtains the total spectrum simultaneously, with photographic detection as the initial method used both historically in the laboratory and in the initial sounding rocket measurements. At this time, arrays of pixels are beginning to be widely used at the focal plane in place of photographic film or plates.

The spectrometers flown on the S3-4 satellite, *Huffman et al.*, 1979, 1980, are shown schematically in Figure 4.2. The figure shows one of two very similar one-quarter meter Ebert-Fastie spectrometers, with the VUV unit covering approximately 107 to 193 nm and the UV unit covering approximately 162 to 290 nm. The VUV detector is an EMR 542G-08 photomultiplier with a cesium fluoride photocathode and a lithium fluoride window. The UV detector is an EMR 542F-09 photomultiplier with a cesium telluride photocathode and a lithium fluoride window. Both wavelength ranges are scanned repetitively with a 22 second scan time. The entrance and exit slit widths can be set to either 2.5, 0.5, or 0.1 nm bandwidths by command. The photomultipliers were operated in the counting mode. The sensor is very similar to the UV spectrometer on the OGO-4 satellite, *Barth and Mackey*, 1969, except that the photomultipliers were operated in the current mode on that sensor.

The sensitivity, S_B, of the spectrometers at 2.5 nm bandwidth is 56 counts/Rayleigh-second at 121.6 nm (HLyα) and 40 counts/Rayleigh-second at 250 nm. The sensitivity varies with wavelength, primarily due to the changing quantum efficiency of the photocathode of the photomultiplier. The sensitivity is obtained from Eq. 4.1 in a manner similar to the photometer, as discussed in the last section.

The values of A and Ω in Eq. 4.1 are determined by the geometry of the spectrometer, in the absence of a telescope. The aperture A is the area of the entrance slit, and the value of Ω is determined by the focal length of the collimating mirror and the size of the ruled area of the grating. The field of the spectrometer in Figure 4.1 is an f/5 square, leading to an Ω of 0.04 steradians. If a matched telescope is used, the aperture and FOV of the telescope can be used in Eq. 4.1. To avoid losing efficiency, the throughput, or AΩ of the telescope should match the throughput of the spectrometer without the telescope. It is best to measure the field of view and the out-of-field rejection of the complete sensor rather than to depend on the specifications used for fabrication.

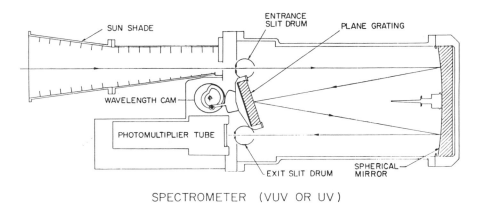

Figure 4.2: Ebert-Fastie spectrometer as flown on S3-4 satellite, *Huffman et al.*, 1980

The value of E in counts per photon, as needed for Eq. 4.1, is obtained from a radiometric calibration, as described in Section 4.5. This calibration for the S3-4 sensors is shown in Figure 4.3. The counts per photon are then combined with the measured field of view and aperture to give the sensitivity.

SPECTROMETER OPTICS

Many spectrometer optical designs have been used over the years, but the most popular are the following.

The **Ebert-Fastie**, or **Ebert**, spectrometer design was developed by *Fastie*, 1963, and coworkers, into a rocket and satellite spectrometer design well suited for obtaining auroral and airglow spectra. As shown in Figure 4.2, the entrance slit is placed at the focus of a collimating mirror, which reflects a collimated beam toward the plane diffraction grating. The light diffracted by the grating is then refocused on the exit slit and detected. The only mechanical motion required to scan the spectrum is to rotate the grating, so that different wavelengths are passed through the exit slit for detection. As the entrance and exit slits are off the optic axis, the wavelength resolution is found to be improved by the use of curved slits, but this refinement is not always necessary. There are a number of designs similar to the Ebert mount, including the Czerny-Turner design, which uses two separate, smaller mirrors rather than the one mirror shown in Figure 4.2. The use of this mount is

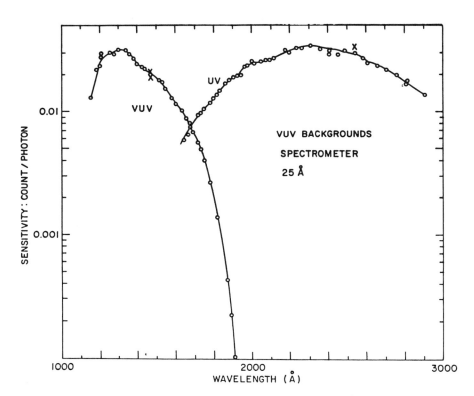

Figure 4.3: Radiometric calibration of the S3-4 spectrometers, *Huffman et al.*, 1980

generally at wavelengths longer than 110 nm, due to the number of reflections needed.

The focal length of the mirror in the S3-4 spectrometer shown in Figure 4.2 is one-fourth meter. The one-eighth meter version is also very useful, having been used for many rocket and satellite measurements. One example is the Horizon Ultraviolet Program (HUP) measurement of the earth limb radiance from the shuttle, which is discussed in Chapters 11 and 12. They also are used in the auroral imagers on the HILAT and Polar BEAR satellites, to be described in Chapter 16. The one-fourth and one-eighth meter Ebert spectrometers used for S3-4, HUP, HILAT, and Polar BEAR were made by Research Support Instruments, Inc., Cockysville, Maryland.

The **Wadsworth** mount is becoming popular as the basis for UV spectrographs with array detectors. In this mount, collimated light incident on a concave grating is found to have stigmatic images near the normal to the grating. These stigmatic images, over a reasonable focal plane, are well suited for detection by a one or two-dimensional array. With a two-dimensional array and the stigmatic image, it will be possible to have the wavelength distribution in one direction and spatial information, such as the variation of radiance in the earth limb, along the other direction. Array detectors will be discussed in more detail in the next section. The Ebert spectrometer field is not as suitable for this purpose as the Wadsworth, as the image is astigmatic and the field is less planar. The incident beam may be mechanically collimated in the EUV.

Finally, the **Rowland circle** mount should be mentioned, although spectrometers and spectrographs based on this principle are more likely to be found in laboratory instruments rather than flight sensors. A concave grating, an entrance slit, and either an exit slit or a focal plane detector placed on a circle with diameter equal to the radius of curvature of the concave grating will be found to have the entrance slit imaged along the circle. This circle is called the Rowland circle. Both near normal and grazing incidence angles can be used. As there is only one reflection, this mount is extremely useful in the ultraviolet, where reflectivities may be poor.

DIFFRACTION GRATINGS

Diffraction gratings are now available in a wide variety of grooves per millimeter and ruled areas. They may be plane or concave, and the method of ruling can be mechanical or holographic. The grating can be coated to improve its efficiency at a given wavelength region. The gratings used can be

4.4. UV IMAGER

either originals or replicas made from a master by making plastic imprints of the ruled surface. Replicas are virtually identical in performance for our purposes and are much cheaper.

There does not appear to be a major reason to choose either *mechanically ruled* gratings, which can have the rulings "blazed", or shaped, to improve the efficiency at a given wavelength range, or *holographic* gratings, which do not form the "ghosts" sometimes found in mechanically ruled gratings due to periodic imperfections in the rulings.

An article on diffraction gratings and mountings by *Hunter*, 1985, is recommended for more detail.

4.4 UV imager

A ground-based imager based on the use of microchannel plates (MCP) for intensification and a charge coupled device (CCD) for detection is shown in Figure 4.4. This imager has been used by C. G. Stergis and coworkers at the Geophysics Directorate, Phillips Laboratory (formerly Air Force Geophysics Laboratory). It has been described by *Lowrance et al.*, 1986, and it can be considered representative of a number of imaging systems based on the MCP-CCD combination used both on the ground and in space.

In Figure 4.4, the scene as imaged on the photocathode in UV light is converted to photoelectrons which are amplified by secondary emission in the MCP in a manner similar to that described for the photomultiplier. The MCP is a bundle of many small, hollow, circular tubes, or channels, made of semiconducting material and having several thousand volts potential along the tubes. A typical channel diameter is 20 micrometers, and standard diameters of the total plate are 18, 25, and 40 millimeters. The MCP may have curved channels or be in several sections at different angles to the optic axis, in order to reduce gain-limiting ion feedback. Gains of up to 10^3 per individual MCP stage are typical. Conceptually, the MCP can be considered as a large number of tiny photomultiplier tubes.

Following amplification by the MCP, the electrons strike a phosphor converting them to visible wavelengths. The MCP-phosphor combination is sometimes called an image intensifier. The visible image is then placed on a CCD having several hundred elements in each direction. It is also usually smaller than the intensified image. The image is shrunk either with a fiber optics coupler, as shown in Figure 4.4, or with a lens. The visible image on the CCD is then converted to a video signal at frame rates of about 30 per

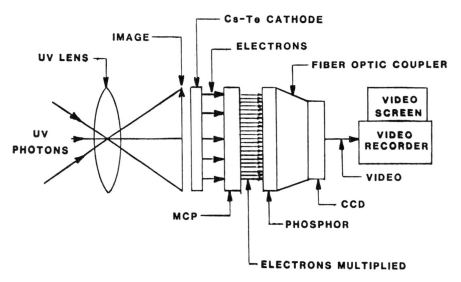

Figure 4.4: UV imager using microchannel plate and CCD (Figure from C. G. Stergis, Phillips Laboratory.)

4.4. UV IMAGER

second, which can be recorded and displayed. The signal produced by the CCD is generally an analog current, but at some point in the data processing it may be digitized and become a digital signal. This digital output can be considered a count-equivalent for comparison with other sensors.

The imager sensitivity in counts/Rayleigh-second can be described with Eq. 4.1, but E, or the number of counts per photon, will usually be less than for photometers and spectrometers. One imager using similar technology has a peak E of 0.003 counts per photon.

The solid angle Ω can be either for the total scene or for individual pixels. The term *pixel*, for picture element, is widely used to designate the number of elements composing a scene or the number of separate detectors in an array. The image size in effective pixels may be smaller than the CCD array size, because of imperfections in the different stages in the imaging process. Thus, the effective pixel size must be found by calibration and test of the imager, or from examination of the image obtained.

Imagers of the type shown in Figure 4.4 can be used in space sensors. They can also be used throughout the UV with changes in imaging method, photocathode material, and filters. The photocathode may be a semitransparent coating on a window or it may be deposited on the face of the MCP. The uncoated MCP has an acceptable quantum efficiency for EUV photons.

DETECTORS FOR IMAGING

Detection methods for the UV focal plane are under very active development, as reviewed by *Timothy*, 1983, and by *Allington-Smith and Schwartz*, 1984. There are now several methods available, with a total pixel count of several million possible. Most methods can be used as one-dimensional, or linear, arrays or as two-dimensional, or area, arrays. The principal methods in use include the following:

Independent element arrays include the *Digicon*, the *Reticon*, and other similar detectors. The Digicon uses an array of separate anodes that receive the electrons after magnetic focusing or after acceleration from a photocathode in close proximity to the anodes, *Beaver et al.*, 1972; *Russak et al.*, 1979. The principal limitation is that separate counting electronics must be provided for each anode, so the array size is limited to several hundred. This limitation did not prevent its selection for use on the first generation of the Hubble Space Telescope. The Reticon and similar devices use a self-scanned linear array of diodes. These have become widely used as spectrometer detectors.

Charged coupled device arrays can be used directly as UV detectors, if the absorption properties of the materials used can be decreased through the use of thin layers, *Janesick et al.*, 1986. This work is continuing, using many approaches that are beyond our scope to review here, but direct UV detection with CCDs are still not widely used. In the UV imager described in this section, the CCD is actually a visible light detector, as the image is intensified and converted to visible wavelengths by the phosphor. This is called an **intensified CCD** detector. The most frequently heard difficulty with these and any current measuring array is the limited dynamic range for a given gain setting.

Position coding arrays are one of the most significant recent developments for array detectors. In each cycle of operation, the position in the field of a single electron cloud from a MCP is coded, or located. This is done by a grid or a group of elements that sense the location of the charge pulse by use of a relatively small number of detector electronic circuits. With fast electronics, an image is rapidly built up of the scene at the photocathode of the MCP. This approach is especially suitable for astronomy applications. The **MAMA**, or Multi-Anode Microchannel Array, *Timothy and Bybee*, 1986, is available in many versions and sizes, *Timothy et al.*, 1989. A MAMA tube having 2000 by 2000 pixels on each side is planned for a second generation Hubble space telescope detector. The **wedge and strip** approach uses the smallest number of electronic circuits for location purposes, and it is adequate for many spectroscopic needs; see, for example, *Siegmund et al.*, 1986. The number of elements can be as small as three. By measuring the arrival time of electron pulses from these three wedge and strip shaped anodes, the position of the arriving charge pulse can be found. There have been a number of arrays based on resistive anodes, including the Codacon and the Ranicon.

Video image tubes have been used in space, with the most successful being possibly the imager on the UV astronomy satellite International Ultraviolet Explorer (IUE), *Bogess et al.*, 1978a, 1978b. They are not used much for remote sensing, possibly because of their perceived complexity, relative to the array type sensors now available.

Electrography is the electron analog of photography with photons, using special film to record the images of photoelectrons produced by

4.5. LABORATORY AND SPACE CALIBRATION

UV photons imaged on a photocathode. Electrography has been long used, *Carruthers*, 1969, 1986, particularly for UV astronomy, but it has typically required the return of the film from space for development, which is a limitation.

4.5 Laboratory and space calibration

Sensors intended for space flight need to have a radiometric calibration and other tests to demonstrate the field of view and out-of-band response. These are usually done together with environmental tests that are described in the next chapter.

Ultraviolet calibration methods and services have been developed extensively, and the problems of agreement between laboratories due simply to calibration uncertainties found in earlier years have been greatly reduced. All UV experimenters should be aware of the work of the National Institute of Standards and Technology (NIST), Gaithersburg, Maryland, (formerly the National Bureau of Standards, or NBS). Standard sources and detectors and the services of the SURF II synchrotron light source are available. The current approaches used are described by *Ott et al.*, 1986, *Canfield and Swanson*, 1987, and references therein. The standard diode detector described in the latter reference has been found to be very useful in the author's laboratory.

The laboratory calibration of space sensors is greatly improved through use of a calibration facility. The facility in use at the Geophysics Directorate, Phillips Laboratory (formerly Air Force Geophysics Laboratory) is shown in Figure 4.5. Ultraviolet radiation from a monochromator or a point light source is collimated and then enters the aperture of the sensor. The sensor is on a gimbal mount that can be moved in a controlled manner over any angle up to about 90° from the beam. In this way, the field of view (FOV), its uniformity, and the out-of-field response can be measured. A beam probe, consisting of a photomultiplier calibrated against NIST standard diodes, can measure the flux in $photons/sec - cm^2$ for radiometric calibration. The tank can be placed under a vacuum in the low 10^{-6} Torr range for work in the FUV and EUV.

A flight sensor will generally have thermocouples and voltage monitors that will allow status, or "housekeeping" information to be obtained in flight. It is also generally useful to have a small UV test lamp incorporated in the sensor. Operation of this source allows an "aliveness" check beginning with

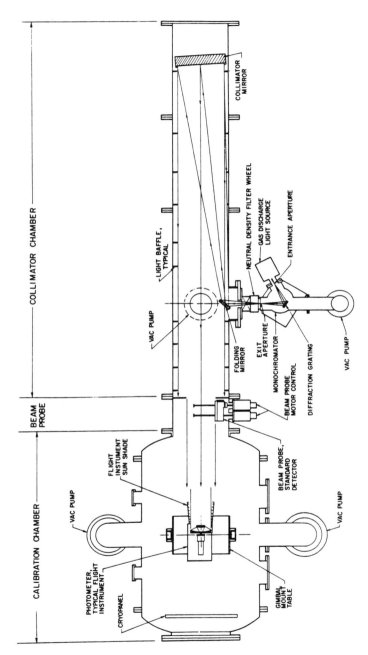

Figure 4.5: Ultraviolet calibration facility at Phillips Laboratory

4.5. LABORATORY AND SPACE CALIBRATION

photons that is useful to have throughout the preflight tests. It also serves as a rough indicator of changes in calibration in flight.

Associated with the sensor, and generally developed with it, should be ground support equipment, or GSE. Ideally, this instrumentation should simulate the spacecraft power, command, and data interfaces for testing before and during integration and before launch. Tests made with the GSE are important to assess the sensor's planned operation in space.

On-orbit calibration appears to be almost essential for the best UV measurements on satellites. It has been found that lower altitude UV sensors in particular seem to degrade in sensitivity when in orbit. This may be due to window and detector *contamination* from outgassing materials on the spacecraft, especially for the early experiments before the most suitable materials for space use were known. Much has been learned about materials in space from the many satellite and shuttle flights that have been conducted to date. Even with these materials, nothing can replace extreme care in practice to avoid contamination, as described by *Hunter*, 1977.

Particle radiation found in space no doubt is a contributor to the degradation of detector performance in space. Shielding must be provided to alleviate, if not eliminate, this problem. Some materials such as sapphire are avoided in space instrumentation because of previously found scintillations induced by space radiation, which adds to the dark count rate of the sensor.

Atomic oxygen in the atmosphere may rapidly destroy materials in the low orbits of less than 300 km used by the space shuttle. Atomic oxygen has been found to rapidly remove a reflective rhodium coating placed on a grating, and reaction products with other materials may redeposit on optical materials. Again, approved materials lists have been developed and should be consulted.

Flight standard sources can be used for measurements on the shuttle, major satellites such as the UARS, and future space stations. A series of deuterium standard lamps have been used by SUSIM, a solar irradiance monitor on the shuttle, *VanHoosier*, 1988. Ionization chambers filled with rare gases are being developed as a standard detector in space, *Carlson, et al.*, 1984.

The **sun** has been used as a calibration source for the ASSI/San Marcos satellite (see Chapter 5). Measurements of the solar flux during the flight of the satellite were done with the airglow sensors, after accurately attenuating the much more intense solar radiation.

Ultraviolet stars offer a very powerful method to solve the on-orbit calibration problem, and it is likely that this method will become quite popular in the future. As the result of measurements by a number of UV astronomy satellites, including the International Ultraviolet Explorer (IUE), the Orbiting Astronomical Observatory program, Copernicus, TD-1, and also from measurements by sounding rockets, the absolute intensity as a function of wavelength is known for a number of stars.

At the present time, the use of UV stars is feasible down to a lower wavelength of about 110 nm in the far ultraviolet. An atlas of the spectra of 143 stars from the IUE program is available, *Wu et al.*, 1991, and similar catalogs can be found for the earlier programs. Much remains to be learned from all-sky surveys that will also be applicable to atmospheric sensing. The EUV is scheduled for a similar comprehensive measurement program on several planned satellites, including the EUVE (Extreme UV Explorer) and the Lyman-FUSE (Far Ultraviolet Spectroscopic Explorer). The recently launched ROSAT satellite covers part of our wavelength range. The two telescopes have ranges of 0.06 to 10 nm and 6 to 30 nm, and one of the primary goals is an all-sky survey.

Lists of stars selected as the in-flight calibration sources for the Hubble Space Telescope have been published, *Bohlin et al.*, 1990; *Turnshek et al.*, 1990. The former reference is particularly useful in that it contains the spectra of the 37 stars chosen for their calibration standards. The spectra are based on an average of several measurements in the 115 to 330 nm range. The star calibration procedures adopted for the SOLSTICE, or SOLar-STellar Irradiance Comparison Experiment, a part of the UARS satellite, are described by *Rottman and Woods*, 1988, together with a list of 28 candidate standard stars. These have only a few duplications with the Hubble list.

Standard stars are nonvariable and ideally have been measured by several satellites. The locations of the stars relative to the orbit flown; the intensities as a function of wavelength; and the pointing capabilities, FOV, and sensitivity of the sensor will determine which stars are most suitable for a given experiment. The use of stars in a solar measurement has the complication that on the average the solar flux at the sensor is 10^8 times larger than stellar fluxes. The flux level of the stars is more suitable for comparison with measurements of the airglow and other atmospheric radiance sources.

4.6 References

Allington-Smith, J. R. and H. E. Schwartz, Imaging photon detectors for optical astronomy, *Quart. J. Royal Astron. Soc., 25*, 267–289, 1984.

Barth, C. A. and E. F. Mackey, Ogo-4 ultraviolet airglow spectrometer, *IEEE Trans. Geosci. Electron. GE-7(2)*, 114–119, 1969.

Beaver, E. A., C. E. McIlwain, J. P. Choisser, and W. Wysoczanski, Counting image tube photoelectrons with semiconductor diodes, *Advances in Electronics and Electron Physics, 33B*, L. Marton, editor, Academic Press, p. 863, 1972.

Bogess, A., F. A. Carr, D. C. Evans, D. Fischel, H. R. Freeman, C. F. Fuechsel, D. A. Klinglesmith, V. L. Krueger, G. W. Longanecker, J. V. Moore, E. J. Pyle, F. Rebar, K. O. Sizemore, W. Sparks, A. B. Underhill, H. D. Vitagliano, D. K. West, F. Macchetto, B. Fitton, P. J. Barker, E. Dunford, P. M. Gondhalekar, J. E. Hall, V. A. W. Harrison, M. B. Oliver, M. C. W. Sandford, P. A. Vaughan, A. K. Ward, B. E. Anderson, A. Boskenberg, C. I. Coleman, M. A. J. Snijders, and R. Wilson, The IUE spacecraft and instrumentation, *Nature, 275*, 372–377, 1978a.

Bogess, A., R. C. Bohlin, D. C. Evans, H. R. Freeman, T. R. Gull, S. R. Heap, D. A. Klinglesmith, G. R. Longanecker, W. Sparks, D. K. West, A. V. Holm, P. M. Perry, F. H. Schiffer III, B. E. Turnrose, C.-C. Wu, A. L. Lane, J. L. Linsky, B. D. Savage, P. Benvenuti, A. Cassatella, J. Clavel, A. Heck, F. Macchetto, M. V. Penston, P. L. Selvelli, E. Dunford, P. Gondhalekar, M. B. Oliver, M. C. W. Sandford, D. Strickland, A. Boksenberg, C. I. Coleman, M. A. J. Snijders, and R. Wilson, In-flight performance of the IUE, *Nature, 275*, 377–385, 1978b.

Bohlin, R. C., A. W. Harris, A. V. Holm, and C. Gry, The ultraviolet calibration of the Hubble Space Telescope. IV. Absolute IUE fluxes of Hubble Space Telescope standard stars, *Astrophys. J.. Suppl., 73*, 413–439, 1990.

Canfield, L. R. and N. Swanson, Far ultraviolet detector standards, *J. Res. NBS, 92*, 97–112, 1987.

Carlson, R. W., H. S. Ogawa, E. Phillips, and D. L. Judge, Absolute measurements of the extreme UV solar flux, *Appl. Optics, 23*, 2327–2332, 1984.

Carruthers, G. R., Magnetically focused electronographic image converters for space astronomy applications, *Appl. Optics, 8*, 633–638, 1969.

Carruthers, G. R., The Far UV Cameras (NRL-803) Space Test Program shuttle experiment, *Ultraviolet Technology*, R. E. Huffman, editor, *Proc. SPIE, 687*, 11–27, 1986.

Fastie, W. G., Instrumentation for far-ultraviolet rocket spectroscopy, *J. Quant. Spectrosc. Radiat. Transfer, 3*, 507–518, 1963.

Hennes, J. and L. Dunkelman, Ultraviolet Technology, Chapter 15 in *The Middle Ultraviolet: Its Science and Technology*, A. E. S. Green, Editor, Wiley, 1966.

Huffman, R. E., F. J. LeBlanc, J. C. Larrabee, and D. E. Paulsen, *Satellite Atmospheric Radiance Measurements in the Vacuum Ultraviolet*, AFGL-TR-79-0151, Air Force Geophysics Laboratory, 1979.

Huffman, R. E., F. J. LeBlanc, J. C. Larrabee, and D. E. Paulsen, Satellite vacuum ultraviolet airglow and auroral observations, *J. Geophys. Res., 85*, 2201–2210, 1980.

Hunter, W. R., Optical contamination: Its prevention in the XUV spectrographs flown by the U. S. Naval Research Laboratory in the Apollo Telescope Mount, *Appl. Optics, 16*, 909–916, 1977.

Hunter, W. R., Diffraction gratings and mountings for the vacuum ultraviolet spectral region, Chapter 2 in *Spectrometric Techniques, 4*, G. A. Vanasse, editor, Academic Press, 1985.

Janesick, J., D. Campbell, T. Elliot, T. Daud, and P. Ottley, Flash technology for CCD imaging in the UV, *Ultraviolet Technology*, R. E. Huffman, editor, *Proc. SPIE 687*, 36–55, 1986.

Lowrance, J. L., C. G. Stergis, and D. F. Collins, Image system for the middle UV, *Ultraviolet Technology*, R. E. Huffman, editor, *Proc. SPIE 687*, 136–141, 1986.

Malacara, D., *Geometrical and Instrumental Optics*, Academic Press, 1988.

4.6. REFERENCES

Ott, W. R., L. R. Canfield, S. C. Ebner, L. R. Hughey, and R. P. Madden, XUV radiometric standards at NBS, *X-ray Calibration: Techniques, Sources, and Detectors, Proc. SPIE, 689*, 178–187, 1986.

Rottman, G. J. and T. N. Woods, In-flight calibration of solar irradiance measurements by direct comparison with stellar observations, *Recent Advances in Sensors, Radiometry, and Data Processing for Remote Sensing, Proc. SPIE, 924*, 136–143, 1988.

Russak, S. L., J. C. Flemming, R. E. Huffman, D. E. Paulsen, and J. C. Larrabee, Development of proximity and electrostatically focused diodes (digicons) for UV measurements from sounding rockets, AFGL-TR-79-0006, Air Force Geophysics Laboratory (now Phillips Laboratory), 1979.

Samson, J. A. R., *Techniques of Vacuum Ultraviolet Spectroscopy*, Wiley, 1967.

Siegmund, O. H. W., M. Lampton, J. Bixler, S. Chakrabarti, J. Vallerga, S. Bowyer, and R. F. Malina, Wedge and strip image readout systems for photon counting detectors in space astronomy, *J. Opt. Soc. Am., A, 3*, 2139–2145, 1986.

Smith, W. J., *Modern Optical Engineering, 2nd Edition*, McGraw-Hill, 1990.

Timothy, J. G., Optical detectors for spectroscopy, *Pub. Astron. Soc. Pacific, 95*, 810–834, 1983.

Timothy, J. G. and R. L. Bybee, High-resolution pulse-counting array detectors for imaging and spectroscopy at ultraviolet wavelengths, *Ultraviolet Technology*, R. E. Huffman, editor, *Proc. SPIE 687*, 109–116, 1986.

Timothy, J. G., J. S. Morgan, D. C. Slater, D. B. Kasle, R. L. Bybee, and H. E. Culver, MAMA detector systems: A status report, *Ultraviolet Technology III*, R. E. Huffman, editor, *Proc. SPIE, 1158*, 104–117, 1989.

Turnshek, D. A., R. C. Bohlin, R. L. Williamson, H. O. L. Lupie, J. Koornneef, and D. H. Morgan, An atlas of Hubble Space Telescope photometric, spectrophotometric, and polarimetric calibration objects, *Astrophys. J., 99*, 1243–1261, 1990.

VanHoosier, M. E., Absolute UV irradiance calibration of the solar UV spectral irradiance monitor (SUSIM) instruments, *Ultraviolet Technology II*, R. E. Huffman, editor, *Proc. SPIE, 932*, 291–296, 1988.

Williams, C. S. and O. A. Becklund, *Optics: A Short Course for Engineers and Scientists*, Wiley, 1972.

Wu, C.-C., D. M. Crenshaw, J. H. Blackwell, Jr., D. Wilson-Diaz, F. H. Schiffer III, D. Burstein, M. N. Fanelli, and R. W. O'Connell, *IUE Ultraviolet Spectral Atlas*, IUE NASA Newsletter No. 43, Goddard SFC, 1991.

Zaidel', A. N. and E. Ya. Shreider, *Vacuum Ultraviolet Spectroscopy*, Ann Arbor-Humphrey Science Publishers, 1970 (translation by Z. Lerman, originally published in Russian in 1967).

Chapter 5

Space Operations

This chapter is an introduction to what the principal investigator, co-principal investigator, or experimenter needs to know first about remote sensing experimentation under field conditions. Many of the subjects covered are applicable to any field measurement, but the emphasis is on Earth-orbiting satellites.

Topics covered include UV transmission in the atmosphere, satellites and orbits, integration and test, and unwanted radiation from spacecraft glow, rocket engine plumes, and scattering sources. The final section is a brief review of the major exploratory satellite programs conducted to date. These programs will be referenced repeatedly in future chapters.

We will not discuss in any detail various technological specialties that may be needed in a specific project. These specialties include vacuum engineering, spacecraft environmental testing, clean room technology, space-qualified materials and procedures, thermal control on spacecraft, quality assurance and reliability, telemetry, data retrieval, and data processing.

A general reference to the problems discussed in this chapter is *Houghton et al.*, 1984. A useful introduction to many of the topics in this chapter is provided by the Air University Space Handbook, *Cochran et al.*, 1985.

5.1 UV absorption considerations

Ultraviolet measurements in the atmosphere must first come to terms with transmission as a function of altitude. The choice of ground, aircraft, balloon, sounding rocket, or satellite platform may be driven largely by atmospheric transmission. Other features of the various platforms or vehicles for

UV experimentation, such as various in-situ measurements possible on some vehicles, are discussed later in this book in relation to specific remote sensing applications.

Atmospheric transmission was introduced in Chapter 2 and illustrated in Figure 2.6, which gives the altitude of the atmospheric unit optical depth, or 1/e absorption, for both incoming or outgoing UV radiation in the nadir or zenith direction, respectively.

Considering now this transmission in more detail, Figure 5.1 gives the UV wavelength range accessible by common measurement platforms or vehicles. In the figure, the wavelength limits for 1% transmission at an angle of 45° from the zenith are given for the wavelength range of 400 to 200 nm, or the NUV and MUV as defined in this book. These are based on transmission calculations done with LOWTRAN 7 and similar codes, as discussed more completely in Chapter 13. Also see Chapters 7 and 8.

Ground passive observation of UV radiation is a straightforward extension of ground airglow, auroral, and solar observations. Larger collecting optics and improved tracking capability can usually be obtained relative to the vehicles shown in Figure 5.1. The wavelength range in the UV ends abruptly near 300 km, with the rapid onset of absorption by stratospheric ozone.

A feature sometimes forgotten about atmospheric transmission is not illustrated in Figure 5.1. For ground level paths, and up to a few thousand feet in altitude, it is possible, under favorable conditions, to get useful transmission paths of up to possibly two kilometers to a short wavelength limit of somewhat less than 250 nm. In these cases, the transmission path does not include the stratospheric ozone layer. The near surface particulates and tropospheric ozone will generally limit the transmission, however. The stratospheric ozone layer acts as a screen for solar UV, and, with suitable filtering, solar-blind sensing of the UV is possible. Chapter 13 on UV codes discusses this area a little further.

Aircraft can be instrumented with UV sensors. Notice from Figure 5.1, however, that there is very little extension of the wavelength range observable from the Earth's surface to lower wavelengths for typical KC-135 aircraft altitudes of 30,000 feet, which is approximately 10 km. The U2 can be used at stratospheric altitudes of about 60,000 ft., or 20 km, but even in this case there is little extension in UV range. The wavelength range is set by the altitude distribution of atmospheric ozone.

5.1. UV ABSORPTION CONSIDERATIONS

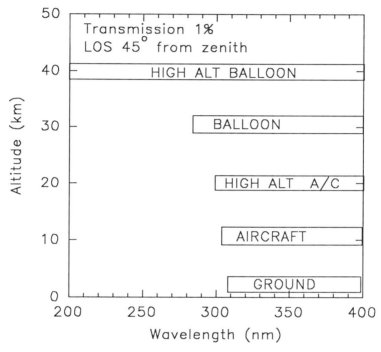

Figure 5.1: Effective UV wavelength ranges for upward viewing from typical observation vehicles; 1% transmission limit shown.

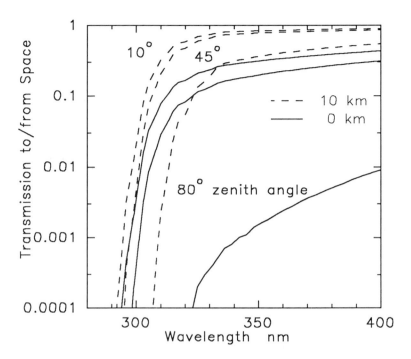

Figure 5.2: Atmospheric transmission from ground level and from a typical aircraft altitude of 10 km, or about 30,000 ft.

Thus, except for the mobility factor, a good ground observation site, with the larger collecting and tracking optics usually possible, may be much superior to an aircraft platform. The aircraft costs are usually quite high, and the effects of vibration and boundary layer turbulence on image resolution in the aircraft case must be evaluated. These considerations may be overridden in any given case by the ability of the aircraft to fly closer to events.

Balloons can typically carry UV payloads of several hundred pounds to altitudes of about 30 km and remain there for hours or days. Smaller payloads can be carried to 40 km for shorter times. Balloons can hover in an area, but they are subject to local winds, which may carry them away from the best observations position. There is a small extension of the wavelength

5.1. UV ABSORPTION CONSIDERATIONS

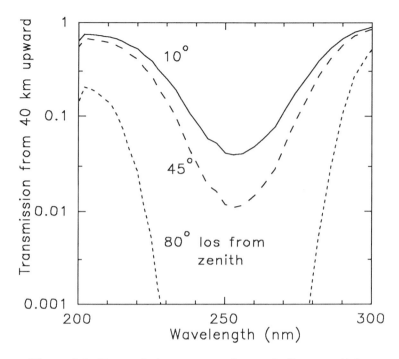

Figure 5.3: Transmission to space from a balloon at 40 km

range toward the short end at 30 km. Significant extension occurs at 40 km, but the far ultraviolet at wavelengths shorter than about 200 nm remains inaccessible. Many slant paths will cause significant transmission losses in the MUV as well. Figure 5.3 gives transmissions to space for a balloon at 40 km.

The transmission calculations in Figures 5.2 and 5.3 were supplied by C. G. Stergis using the PCTRAN 7 version of LOWTRAN 7 described in Chapter 13. For these calculations, the UV Standard Atmosphere 1976 and a rural aerosol model with 23 km visibility were used.

Research aircraft may allow UV experimenters to take advantage of some altitudes difficult to reach otherwise. Altitudes of 75 km flown by the X–15 have been utilized. Planned vehicles such as the National Aerospace Plane (NASP) may offer interesting platforms from which to develop atmospheric remote sensing methods. These platforms will probably not be available for a number of years.

Sounding rockets can provide coverage throughout the UV, but only for time periods of a few minutes. They provide a means to conduct research from many locations around the Earth and in coordination with other observing methods or with events such as eclipses or aurora. Sounding rockets are usually flown to a maximum altitude, or apogee, of 50 to 600 km.

Satellites allow observation of the total UV wavelength range. They orbit at altitudes from about 175 to at least the geosynchronous altitude of approximately 40,000 km, which might be considered an approximate upper altitude limit for useful atmospheric remote sensing in the UV. Satellites give the opportunity to study the UV radiative environment for several years, in the most successful cases. Thus, geophysical effects such as day/night, season, latitude, and solar cycle can be evaluated. Earth orbiting satellites are the most likely platforms for global atmospheric and ionospheric remote sensing systems.

The **Space Transportation System, or shuttle**, flying at an altitude of between about 225 and 300 km, has no transmission limitation for upward viewing. It is, however, beneath significant sources of atmospheric UV radiance, as has been found after the flight as unwanted radiation in some astronomy experiments. For our purposes, the shuttle sortie flights amount to a short-duration (about one week) satellite that allows the use of large and heavy instrumentation, post-flight calibration, and reflight of experiments. Experiments can also be carried on a free-flying satellite launched from the shuttle, which would of course have the typical free-flying satellite lifetime. The presence of astronauts on board may be an advantage, but it is always a complication. Current and foreseeable schedules make shuttle sortie missions a difficult place to do experimentation into remote sensing.

5.2 Satellites and orbits

Satellites and orbits need to be understood, at least in a general way, by the UV experimenter. There are several references available, but two that have been useful to the author are by *King-Hele*, 1960, and by *Davidoff*, 1990. This section is a brief introduction only.

Remote sensing from satellites is illustrated in Figure 5.4, which defines

5.2. SATELLITES AND ORBITS

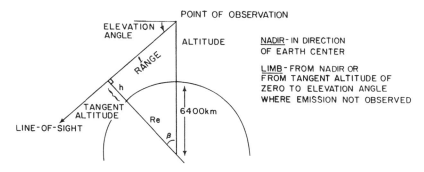

Figure 5.4: Definitions associated with earth viewing from a satellite

several terms. From the point of observation in space, the nadir direction is the direction toward the center of the Earth, and the zenith direction is the opposite. The Earth limb is the region of the atmosphere of interest due to its radiance or absorption properties. These effects are evident for any Earth viewing angle other than nadir, so limb viewing directions are defined here as from nadir to the local horizontal at the observation point. The tangent altitude is the altitude of closest approach of a limb line of sight for directions between the hard Earth edge direction, which would have a tangent altitude of zero, and the local horizontal direction at the observation point, which would have the tangent altitude of the observation point.

All of these terms are found to be useful, even if, for example, transmission considerations prevent any atmospheric radiance generated near the Earth edge from being seen by a UV sensor on a satellite. The Earth limb pointing direction and the tangent altitude are based on geometry. The principal contribution to the observed UV radiance along this line of sight may not be near the tangent altitude, although for the higher tangent altitudes it will be. The simplified approach used here ignores the fact that the earth is not a perfect sphere and that the geodetic field is not uniform, and it should be kept in mind that the determination of the tangent altitude to high accuracy on a global basis from satellites is complicated.

Polar orbiting satellites have been found to be the most useful platforms for UV remote sensing, since the polar region is very important for understanding the atmosphere on a global basis. Low altitude polar orbits have an optimum altitude of about 1000 km, although experimentation at any altitude is useful, with 175 km about the lowest satellite altitudes found. High altitude polar orbits can have apogees up to several Earth radii. Higher altitudes are more costly to reach, and they are limited by the need to shield electronics when passing through belts of space radiation. The data interpretation is simplified with an orbit as nearly circular as possible.

Polar orbiting satellites have inclinations of about 60° or greater. The inclination is the angle between the northward ground track and the eastward direction at the equator, with a 90° inclination passing directly over the poles. The orbit's inclination is the highest latitude reached by the orbit. A commonly used orbit is **sun-synchronous**, arriving at each latitude at the same local time on every orbit. Because of the Earth's oblateness, sun synchronous operation is achieved with a retrograde orbit having an inclination of about 98°, for low Earth orbits.

Examples of current operational satellites in low altitude polar orbit are the DOD Defense Meteorological Satellite Program (DMSP) and the NOAA NIMBUS satellites. The two DMSP satellites that are in flight at the same time have generally been at an altitude of about 830 km. The fixed local times at the equator have usually been 1030 and 2230 ("noon-midnight") and 0600 and 1800 ("dawn-dusk"). These lower altitude polar satellites have periods of about 100 minutes. NIMBUS satellites are usually at a slightly higher altitude and at different local times.

For research purposes, the **non-sun-synchronous** inclinations are preferred, especially if the local time variation is not too rapid. In this way, solar angle effects can be studied. The orbits of the DOD Space Test Program HILAT and Polar BEAR satellites have been of this type.

The **Molniya** type orbit is used for some high altitude remote sensors. Orbits used by the NASA Dynamics Explorer I, or DE I, and the Swedish VIKING are of this type. They are elliptical, with an apogee of several earth radii, a perigee of several hundred kilometers, and a period of several hours. These orbits allow high altitude UV auroral imaging as well as the ability to sample in-situ particles and fields over a range of altitudes in the radiation belts. An attribute of this type of orbit is that the ground track of the near apogee position of the orbit changes in latitude slowly, so that the prime observation time near apogee changes from pole to pole over periods of several months.

Most satellites, the shuttle, and proposed space stations are generally in

5.2. SATELLITES AND ORBITS

nonpolar orbits, having inclinations of less than about 60°. The shuttle and currently proposed space stations are in low Earth orbits of about 250 to 300 km.

The widely used **geostationary**, or geosynchronous, orbit places the satellite in a position fixed relative to a given region on the Earth's surface. The height of the geostationary orbit is about 35,800 km, or between five and six earth radii, and the inclination is low, so that the ground track is near the equator. These orbits are widely used for communications. For enviromental sensing, the NOAA GOES network provides cloud imaging and other data needed for weather prediction. Unfortunately, the view of the polar regions is usually considered to be too limited for applications, and possibly for this reason they have not been used for UV atmospheric remote sensing. As the importance of the equatorial airglow belt becomes better understood, this may change.

Satellite **stabilization and tracking** may be crucial to the ability to sense the atmosphere. An unstabilized satellite would tumble in its orbit with no ability to control the line of sight of sensors. The UV experimenter must be familiar with stabilization methods, but it is a specialized subject and only a few comments will be made. The interested reader is refered to books such as *Wertz*, 1978, for details.

The most common types of stabilization are *three-axis stabilization*, where the position of the satellite relative to the Earth is maintained; *spin stabilization*, where the line of sight can be placed perpendicular to the spin axis so that views that sweep across the Earth disk from space are possible; and *inertial stabilization*, where the directions of some or all of the satellite coordinates are fixed in inertial space, so that the direction relative to the celestial sphere remains constant. In any given case, some of these may be combined. Stabilization can be passive, with momentum wheels and gravity gradient booms, or active, with on board propulsion to adjust the satellite coordinates and lines of sight relative to the pitch, yaw, and roll axes.

Tracking and pointing refer to the ability of the spacecraft, or instrumentation on board, to point at and keep within a specified field of view objects, such as stars, or desired pointing directions, such as a specific tangent altitude, as the spacecraft moves in its orbit. To accomplish stabilization, tracking, and pointing, a range of instrumentation, including IR horizon sensors, star trackers, cold-gas thrusters, and reactive thrusters, are available. Many other factors must be considered, such as jitter, dead band, control system frequencies, etc.

The space experimenter may be constrained in experiment design by capabilities for stabilization, tracking, and pointing. Typically available three axis stabilization is 1° in pitch, yaw, and roll, but much better values to possibly 0.01^o may be available in special cases.

Ultraviolet radiance from the Earth limb has been proposed as a reference for horizon sensing, *Wolff*, 1966, 1967. Measurements with this use in mind have been flown as the Horizon Ultraviolet Program, HUP, experiment, *Huffman et al.*, 1981, 1983. HUP also flew on the shuttle flight STS-39 in 1991.

Scientific length units are usually kilometers, but spacecraft altitudes may be given in nautical miles, which are 1.85 km. The nautical mile may be abbreviated nm, but in this book nm stands for the wavelength in nanometers. A statute mile is 1.61 km, or 5280 feet. As tropospheric and stratospheric altitudes associated with aircraft are frequently given in feet, it is useful to know that a kilometer is 3280.8 feet.

5.3 Integration and test

After a sensor has been designed and built, it must be calibrated and tested both as an individual unit and, during the integration phase, as a part of the total spacecraft. Calibrations in the laboratory and in space were described in Section 4.5. The tests described here are for satellite sensors; a similar but abbreviated series of tests is generally used for sounding rocket experiments.

An **Interface Control Document** (ICD) or similar document should be established early in the program, preferably before the design of the sensor is complete. The ICD defines the interfaces between sensor and spacecraft. It is generally prepared by the spacecraft integrating organization for the principal sponsor of the program. It is an agreement among all parties involved in the program, including the spacecraft integrating organization, funding agencies, and the principal investigator. There are usually several sensors, each with a principal investigator.

An **interface** is a place where the sensor and the spacecraft have to meet properly for the achievement of an objective of a project. The concept usually includes properties of the sensor that must be met for it to be readily integrated into the planned spacecraft. The ICD defines these interfaces. The interfaces usually considered include the following: size; weight; power; thermal control; electromagnetic interference; vibration and shock; mounting and field of view; contamination control; approved materials, com-

5.3. INTEGRATION AND TEST

ponents, and procedures; data formats and processing; number and type of commands; on-orbit operating procedures; status sensors; and any special requirements. As the mission plans develop, it is essential that the ICD be maintained as a controlled document, with changes agreed to or at least understood by all parties.

Environmental tests, as defined in some detail by the ICD, must be passed by the sensor before integration and before flight. These vary with the program, spacecraft, and philosophy of the spacecraft integrator and program sponsor. Sometimes these are conducted on protoflight sensors, but in many cases, the tests are on the flight article. In the latter cases, care is needed to be sure that the sensor is not damaged or degraded in the testing process. The following tests are commonly done:

Thermal The sensor must operate over the planned temperature range of the spacecraft, and nonoperating and survival temperature limits must be found or estimated.

Thermal-vacuum This test subjects the sensor to the thermal and vacuum environments expected in the flight, plus the gradients expected between the high and low temperatures. A series of tests over hours or days is used to demonstrate this behavior.

Vibration The vibrational load including frequencies and accelerations associated with the launch and, in the case of the shuttle, the landing of the spacecraft. There may also be shock test requirements. Vibration tests are usually done in all three axes.

Electromagnetic interference The EMI tests must demonstrate that the sensor is not sensitive to the frequencies and EM radiation levels specified for the spacecraft, and also that it does not radiate EM signals that could cause interference to other equipment on the satellite.

Most of these environmental tests will generally be performed on the sensor alone before delivery of the sensor to the integrating contractor. These tests are generally combined with the radiometric, geometric, and spectral calibrations required to obtain valid measurements, as discussed in the previous chapter. Delivery to the spacecraft integrating contractor of the flight-worthy sensor and the associated ground support equipment generally occurs at least one year before launch. The experiment team will then support the integration tests, but under the control of the integrating contractor.

Pressure to deliver sensors without adequate calibration and testing must be resisted by the principal investigator and team. Knowledgeable program management will insist upon complete tests, calibration, and documentation before flight. However, other factors such as the budget and flight opportunities have to be considered. While the long term quality of the measurement is the primary concern of the principal investigator, schedules that have been previously negotiated cannot be extended indefinitely.

After successful completion of the integration phase, the spacecraft is taken to the launch range, mounted on the booster, and launched. Final preflight tests using the ground support equipment developed both for the sensors and for the spacecraft are generally done as close to launch time as practicable. The on-orbit phase begins after initial checks and out-gassing in orbit are finished, which may take several weeks. Then, useful data can be acquired by the sensor.

On-orbit operations may well extend over years and lead to the acquisition of a sizable amount of data. The reduction, analysis, and interpretation of these measurements must be included in the total experiment plan, but details will vary with the specific program.

5.4 Glows and plumes

The emissions from spacecraft and their operation are of interest by themselves and also as possible optical contaminants that will degrade the performance of sensors. These potential optical contaminants can be either passive emissions, such as shuttle glow, or active sources, such as rocket exhaust plumes.

SPACECRAFT GLOW

Visible light emission, or glow, from the surfaces of the shuttle was found in the early flights, and it has excited much interest. A popular account of shuttle glow has been given by *Hunton*, 1989. The primary contribution to the visible emission appears to be a variant of the well-known day airglow continuum emission from NO_2 due to the reaction of atomic oxygen with nitric oxide, NO, *Paulsen, et al.*, 1970. The mechanism remains poorly understood, but it almost certainly involves heterogeneous reactions that lead to the formation of nitric oxide and its subsequent emission. This emission, due to reaction with atomic oxygen, is well known in both laboratory and

5.4. GLOWS AND PLUMES

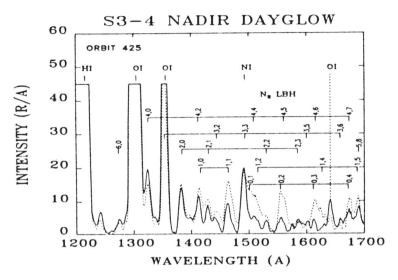

Figure 5.5: UV spacecraft glow (dotted lines) and airglow (solid lines) (*Conway et al.*, 1987)

atmosphere. It is sometimes called the airglow continuum, as it appears to be a continuum with very diffuse bands superimposed. It usually does not extend much below 400 nm, which is near the bond energy of an N-O bond in NO_2. It has not been demonstrated to date to occur at ultraviolet wavelengths, and thus the shuttle glow is not known to be within our scope.

Ultraviolet spacecraft glow is different from shuttle glow. It was first found by the spectrometers and photometer on the S3-4 satellite, *Huffman et al.*, 1980. This emission was identified as due to the Lyman-Birge-Hopfield bands of molecular nitrogen but with a vibrational distribution different from that found in the day airglow and in the aurora. The initial analysis of the bands, *Meier and Conway*, 1983, followed the first report in assuming that the emission was an airglow.

Further investigation of the altitude dependence of the emission indicates that the emission can be explained as a spacecraft glow, *Conway et al.*, 1987. The intensity varies with the third power of the density, however, so it becomes very weak at altitudes above 200 km. Thus, its appearance is associated with the very low altitude of this satellite during part of its orbit.

A search for this type of LBH emission on the shuttle has found it to occur very weakly, *Torr et al.*, 1985.

Ultraviolet and visible spacecraft glow are related to several other types of emission associated with the high velocity of the spacecraft passing through the atmosphere. These additional emissions may be found on reentry into the atmosphere, as a part of the bow shock from high velocity boosters, and by release of chemicals in the upper atmosphere from a high velocity vehicle. These topics are outside our scope except as the emission mechanisms might be similar to a spacecraft glow.

Spacecraft glow appears to involve heterogeneous reactions, atomic oxygen, and the kinetic energy somehow imparted to the emission reaction from the spacecraft velocity. Possible mechanisms for the ultraviolet spacecraft glow can be found in *Kofsky*, 1988; *Cuthbertson and Langer*, 1989; *Meyerott and Swenson*, 1991, and references therein. There is need for controlled experimentation both in space and in the laboratory to fully understand the UV spacecraft glow.

The glow discovered by the S3-4 UV sensors has sometimes been suggested as invalidating the airglow measurements made by this satellite. This is not the case, as airglow measurements uncontaminated by the glow are possible at higher altitudes where the glow cannot be detected. The most recent demonstration of this is the measurement of the weak NO and O_2 nightglow with no trace of the spacecraft glow at the altitudes used, *Eastes et al.*, 1992. The S3-4 sensors are used as illustrative examples in Chapter 4.

Optical contamination around the shuttle has been of particular concern from the early flights, and an initial report of measurements with the ISO instrument is available, *Torr and Torr*, 1985. At this time, it can be said that there have been a number of successful optical measurements from more than forty flights of the shuttle. However, precautions must be taken regarding fields of view and time periods when thrusters are operating or chemicals are being dumped. Release of NO gas has resulted in a large increase in the shuttle glow, *Viereck et al.*, 1991. This is dramatic evidence for the NO_2 emission mechanism mentioned above.

ROCKET EXHAUST PLUMES

The ultraviolet emission from rocket exhaust plumes is a potential optical contaminant for sensors. The shuttle, for example, has many primary control

5.4. GLOWS AND PLUMES

system thrusters which are short duration rocket engines giving optical and UV emission. These are potential problems for optical measurements from the shuttle, although UV radiance measurements without exhaust plume contamination have been made from the shuttle, *Huffman et al.*, 1981, 1983.

The use of the ultraviolet for missile exhaust plume observations was described a number of years ago by *Berlad*, 1966, and by *Yoshihara*, 1966. Flames, including rocket propulsion reactions and chemiluminescence, are described in detail in a book by *Gaydon and Wolfhard*, 1970. The use of the infrared for plume observation is discussed by *Spiro and Schlessinger*, 1989, who describe the growth of exhaust plumes with altitude.

A recent review of our knowledge of the chemiluminescent and other reactions giving UV emission from rocket exhaust plumes has been given by *Kolb et al.*, 1988. It should also be kept in mind that the plume from a rocket in the thermosphere is a gas release that grows to many kilometers in width, *Draper et al.*, 1975. This enormous perturbation of the atmosphere also can lead to radiance due to chemical reactions and scattering.

Cold gas thrusters are used on many spacecraft to change the attitude of vehicles in pitch, yaw, and roll. This cold gas, usually nitrogen, cools and condenses into particles near the thruster. The particle cloud is an excellent scatterer of sunlight, which has lead to unwanted interference in measurements. An account of this scattering by thrusters using argon gas and its elimination by the use of neon for the thruster gas is given by *Kolb et al.*, 1983, 1985. The locations of all thrusters on the spacecraft should be known and planned for by principal investigators.

Sounding rocket solid propellants may be a possible emission source, as has been identified in some experiments from space. The solid propellant does not stop thrusting immediately, as is the case with liquid propellants that have cutoff valves. The spent motor case continues to smolder and release burning pieces of the fuel for some time after "burnout". This emission may be in the FOV of infrared and ultraviolet sensors carried by the payload of the sounding rocket, *Price et al.*, 1980; *Frankel et al.*, 1985.

The ultraviolet experimenter should keep these potential sources of optical contamination in mind in the design and operation of sensors in space. The planning and integration phases of the program are crucial in this regard. A clear, unobstructed FOV is called for, but also the principal investigator should try to obtain a situation where there are no structures extending from the body of the spacecraft, such as a boom or an antenna, which may scatter sunlight that will enter the sensor aperture. Continuous monitoring is required to prevent last minute additions of structures which may lead to degraded performance in orbit due to increased solar scattering.

5.5 Major satellite programs

This section provides a short introduction and reference to major Earth orbiting satellite programs discussed more fully elsewhere in this book. The list is limited to programs that have provided or will provide a significant flight data base. It is also limited to exploratory, or multipurpose, satellite programs whose research goals include understanding the UV radiative environment of the Earth's atmosphere. Therefore, dedicated, or operational, satellites are not included here, but many are mentioned in connection with specific applications. Programs whose primary objectives were astronomy, solar observations, and study of the other planets have also been excluded.

Note that the experiment tends to be given the name of the spacecraft program after some time, but this practice usually causes no problem. Only the UV sensors in the satellite payload are of concern here, and of course many additional measurements from programs and experiments not in the following list are also discussed in this book.

Orbiting geophysical observatory-4 OGO-4 was launched in 1967 into a polar orbit with a dual UV spectrometer and a set of ionization chambers observing in the nadir direction. The observations are primarily discussed in Sections 9.2, 9.3, 10.2, and 16.1.

S3-4 Launched in 1978, this satellite carried the experiment VUV Backgrounds, which consisted of a dual UV spectrometer and a multispectral photometer, into a polar orbit. These sensors are used as illustrations in Chapter 4. The measurements are included in Sections 5.4, 9.2, 9.3, 10.2, 11.1, 12.1, 12.2, 12.3, 13.3, and 16.1.

P78-1 This satellite, launched in 1979, obtained EUV and FUV spectra in a polar orbit. Further discussion is principally in 9.2, 9.3, 10.2, 11.2, and 16.1.

Solar mesospheric explorer SME, launched in 1981, obtained limb radiance measurements from a polar orbit primarily in the MUV. The sensors and measurements are discussed in 11.3, 14.3, and 15.2.

Dynamics Explorer I DE-I was launched in 1982 with several global imagers, including coverage of the FUV using filters. A polar Molniya orbit was used, and an extremely large data base was obtained. It is described in some detail in 16.2 and also in 10.3.

Imaging spectrometric observatory ISO was a part of the Spacelab I mission of the shuttle flown in 1983. This major shuttle sensor, (*Torr, et al.*, 1982), provides spectra and limb imaging over a wide wavelength range including the UV. Measurements are discussed in 5.4.

HILAT and Polar BEAR These similar satellites, launched in 1983 and 1986, respectively, imaged the northern auroral region from a low altitude polar orbit with a modified FUV spectrometer. They are discussed primarily in 16.2, and are included in 2.2, 10.3, 12.3, 17.2, 17.4, and 17.5.

Viking This satellite was launched into a polar Molniya orbit in 1986. It carried two UV imagers with FUV coverage utilizing filters and CCD detectors. They were used primarily for auroral imaging. It is described in some detail in 16.2.

ASSI/San Marco This satellite carried an atmospheric and solar experiment flown in a near equatorial orbit in 1988, with spectrometer coverage of the 20 to 700 nm wavelength range. The measurements are being analyzed and cross calibrated in some detail before release. The sensor has been described, *Schmidtke et al.*, 1985.

5.6 References

Berlad, A. L., Radiation from flames and chemical perturbation of the atmosphere, Chapter 12 in *The Middle Ultraviolet: Its Science and Technology*, A. E. S. Green, Editor, Wiley, 1966.

Cochran, C. D., D. M. Gorman, and J. D. Dumoulin, *Space Handbook*, AU-18, Air University Press, Maxwell AFB, AL, (available through Supt. Documents) 1985.

Conway, R. R., R. R. Meier, D. F. Strobel, and R. E. Huffman, The far ultraviolet glow of the S3-4 satellite, *Geophys. Res. Lett., 14*, 628–631, 1987.

Cuthbertson, J. W. and W. D. Langer, A surface chemistry model for the altitude dependence of the N_2 Lyman-Birge-Hopfield glow on spacecraft, *J. Geophys. Res., 94*, 9149–9154, 1989.

Davidoff, M., *The Satellite Experimenter's Handbook*, The American Radio Relay League, 225 Main Street, Newington, CT 06111, 1990 (written for radio amateurs but useful as an introduction).

Draper, J. S., F. Bien, R. E. Huffman, and D. E. Paulsen, Rocket plumes in the thermosphere, *AIAA Journal, 13*, 825–827, 1975.

Eastes, R. W., R. E. Huffman, and F. J. LeBlanc, NO and O_2 ultraviolet nightglow and spacecraft glow from the S3-4 satellite, *Planet. Spa. Sci.*, to be published, 1992.

Frankel, D. S., M. E. Gersh, A. McIntyre, R. E. Huffman, and D. E. Paulsen, Aries rocket motor infrared and ultraviolet spent-stage emission, *J. Spacecraft and Rockets, 22*, 567–573, 1985.

Gaydon, A. G. and H. G. Wolfhard, *Flames, their structure, radiation, and temperature*, Third ed. rev., Chapman and Hall, 1970.

Houghton, J. T., F. W. Taylor, and C. D. Rodgers, *Remote Sounding of Atmospheres*, Cambridge U. Press, 1984.

Huffman, R. E., F. J. LeBlanc, J. C. Larrabee, and D. E. Paulsen, Satellite vacuum ultraviolet airglow and auroral observations, *J. Geophys. Res. 85*, 2201–2215, 1980.

Huffman, R. E., F. J. LeBlanc, D. E. Paulsen, and J. C. Larrabee, Ultraviolet horizon sensing from space, *Shuttle Pointing of Electro-Optical Experiments*, W. Jerkovsky, editor, *Proc. SPIE, 265*, 290–294, 1981.

Huffman, R. E., F. J. LeBlanc, J. C. Larrabee, D. E. Paulsen, and V. C. Baisley, Ultraviolet horizon radiance measurements from shuttle, *AIAA Shuttle Environment and Operations Meeting*, Washington, D.C., 31 Oct.–2 Nov. 1983, Paper AIAA-83-2628-CP, 1983.

Hunton, D. F., Shuttle Glow, *Scientific American*, 92–98, November, 1989.

King-Hele, D., *Satellites and Scientific Research*, Routldege and Kegan Paul, London, 1960 (US publisher, Dover).

Kofsky, I. L., Excitation of N_2 Lyman-Birge-Hopfield bands emission by low earth orbiting spacecraft, *Geophys. Res. Let., 15*, 241–244, 1988.

5.6. REFERENCES

Kolb, C. E., R. B. Lyons, J. B. Elgin, R. E. Huffman, D. E. Paulsen, and A. McIntyre, Scattered visible and ultraviolet solar radiation from condensed attitude control jet plumes, *J. Spacecraft and Rockets, 20*, 383–389, 1983.

Kolb, C. E., R. B. Lyons, R. E. Huffman, D. E. Paulsen, and A. McIntyre, Mitigation of scattered solar radiation from condensed attitude control jet plumes, *J. Spacecraft and Rockets, 22*, 215–216, 1985.

Kolb, C. E., S. B. Ryali, and J. C. Wormhoudt, The chemical physics of ultraviolet rocket plume signatures, *Ultraviolet Technology II*, R. E. Huffman, editor, *Proc. SPIE, 932*, 2–23, 1988.

Meier, R. R. and R. R. Conway, On the N_2 Lyman-Birge-Hopfield band nightglow, *J. Geophys. Res., 88*, 4929–4934, 1983.

Meyerott, R. E. and G. R. Swenson, N_2 spacecraft glows from $N(^4S)$ recombination, *Planet. Spa. Sci., 39*, 469–478, 1991, and previous papers by these authors.

Paulsen, D. E., W. F. Sheridan, and R. E. Huffman, Thermal and recombination emission of NO_2, *J. Chem. Phys., 53*, 647, 1970.

Price, S. D., T.L.Murdock, A.McIntyre, R. E. Huffman, and D. E. Paulsen, On the diffuse cosmic ultraviolet background measured from Aries A-8, *Astrophys. J., 240*, L1–L2, 1980.

Schmidtke, G., P. Seidl, and C. Wita, Airglow-solar spectrometer instrument (20–700 nm) aboard the San Marcos D/L satellite, *Applied Opt., 24*, 3206–3213, 1985.

Spiro, I. J. and M. Schlessinger, *Infrared Technology Fundamentals*, Dekker, 1989.

Torr, M. R. and D. G. Torr, A preliminary spectroscopic assessment of the Spacelab I/Shuttle optical environment, *J. Geophys. Res., 90*, 1683–1690, 1985.

Torr, M. R., R. W. Basedow, and D. G. Torr, Spectroscopic imaging of the thermosphere from the space shuttle, *Applied Opt., 21*, 4130–4145, 1982.

Torr, M. R., D. G. Torr, and J. W. Eun, A spectral search for Lyman-Birge-Hopfield nightglow from Spacelab I, *J. Geophys. Res., 90*, 4427–4433, 1985.

Wertz, J. R., editor, *Spacecraft Attitude Determination and Control*, Reidel, 1978.

Wolff, M., Profiles of a planetary limb for space navigation, *J. Spacecraft, 3*, 538–542, 1966.

Wolff, M., Precision limb profiles for navigation and research, *J. Spacecraft, 4*, 978–983, 1967.

Viereck, R. A., E. Murad, B. D. Green, P. Joshi, C. P. Pike, R. Hieb, and G. Harbaugh, Origin of the shuttle glow, *Nature, 354*, 48-50, 1991.

Yoshihara, H., Gasdynamics of rocket exhaust plumes, Chapter 13 in *The Middle Ultraviolet: Its Science and Technology*, A. E. S. Green, Editor, Wiley, 1966.

Chapter 6

The Earth's Atmosphere

The envelope of gas surrounding the Earth undergoes many changes as altitude increases. The familiar molecular nitrogen and molecular oxygen at ground level are dissociated and ionized by solar electromagnetic and particle radiation. The pressure and density decrease, the mean free path increases, and the temperature swings back and forth. At the highest altitudes, the atmosphere is virtually fully ionized and thus controlled by particles and fields as it merges with the local solar wind.

It is necessary to have a general knowledge of the atmosphere in order to understand methods of remote sensing using the ultraviolet. Emission and scattering from the upper parts of the atmosphere are used as remote sensing methods for atmospheric density and composition of both neutral or ionized constituents. This chapter is a brief introduction to the atmosphere and to some of its properties.

For reference, the standard temperature and pressure (STP) are 273.15 K and 1013.25 mb, and the corresponding number density is Loschmidt's number, $2.687 \times 10^{19} \, cm^{-3}$. The fractional compositions or mixing ratios for the most abundant constituents in dry, sea level air are as follows: N_2: 0.7808; O_2: 0.2095; Ar: 0.0093; and CO_2: 0.0003. There are many minor species, including the highly variable water vapor.

It must be kept in mind that the atmosphere is dynamic, with changes occurring due to changes in solar illumination, season, latitude, longitude, and solar activity. For a more detailed discussion, see *Brasseur and Solomon*, 1984, and *Warneck*, 1988, for the atmosphere up to about 100 kilometers in altitude, and *Rees*, 1989, for higher altitudes. Standard atmospheres from the surface to 1000 km are discussed by *Champion et al.*, 1985, and the composition is shown in Figures 6.1 and 6.2.

6.1 Atmospheric regions

The atmosphere can be separated into layers on the basis of temperature and composition. For our purposes, it is necessary to know about these regions in a general way. These regions are shown in Figure 6.3.

TEMPERATURE REGIONS are defined in altitude as follows:

Troposphere This region extends from the surface to about 18 kilometers, where there is a temperature minimum of about 210 K called the tropopause.

Stratosphere The temperature rises in the stratosphere, which is between roughly 18 kilometers and the stratopause near 50 kilometers altitude, where a maximum temperature of about 280 K is reached.

Mesosphere The temperature drops again from the stratopause at about 50 kilometers to a minimum just below 200 K at the mesopause near 86 kilometers. This region is called the mesosphere.

Thermosphere The final layer of concern to us is the thermosphere, where the temperature rises continuously from the low value of the mesopause near 86 kilometers. The upper limit is not defined by a temperature reversal but by the place where the temperature becomes constant. This constant temperature can be reached between roughly 500 and 1000 kilometers, with the latter altitude our upper limit of the thermosphere.

Exosphere In terms of temperatures, the exosphere may be considered to begin where the kinetic temperature has risen to become constant. This *exospheric temperature* can be in the range of 500 to 2000 K, and it becomes higher during periods of increased solar activity.

COMPOSITION REGIONS are defined as follows:

Lower atmosphere or homosphere In the homosphere, which includes the troposphere and stratosphere, the relative percentages, or mixing ratios, of the major components N_2 and O_2 remain constant at their surface levels as the pressure and density decrease. This situation prevails up to an altitude of about 86 kilometers (see Figure 6.1).

6.1. ATMOSPHERIC REGIONS

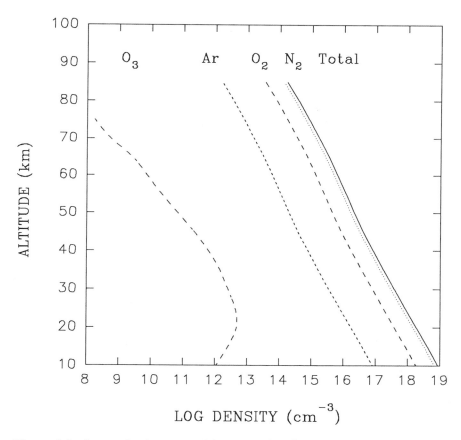

Figure 6.1: Atmospheric composition to 86 km (*U.S. Standard Atmosphere, 1976*)

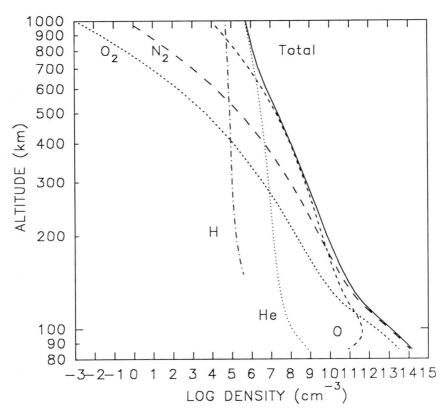

Figure 6.2: Atmospheric composition from 86 to 1000 km (*U.S. Standard Atmosphere, 1976*)

6.1. ATMOSPHERIC REGIONS

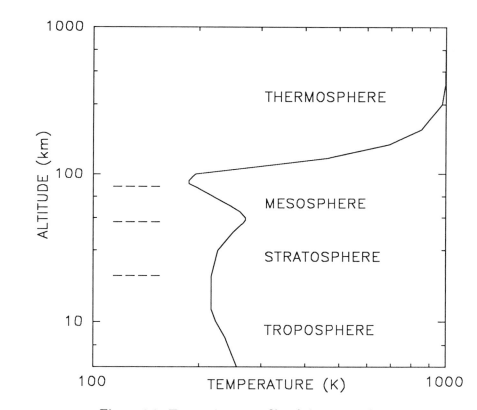

Figure 6.3: Temperature profile of the atmosphere

Upper atmosphere or heterosphere The upper atmosphere, as defined here, includes the mesosphere and the thermosphere. The gross composition and mean molecular weight begin to change near 86 kilometers, as the photodissociation of molecular oxygen and its relatively slow recombination leads to an increasing relative abundance of atomic oxygen. At still higher altitudes, first helium and then atomic hydrogen dominate the composition (see Figure 6.2).

The **IONOSPHERE** is the region of the atmosphere that contains significant numbers of electron and ions, formed initially by photoionization due to solar ultraviolet radiation. These layers or regions are above approximately 70 km. Electron densities in the ionosphere both day and night and for solar maximum and minimum conditions are shown in Figure 6.4.

The names given to the ionospheric layers are:

D-region This layer extends from about 70 to 90 kilometers during the daytime. It is formed from photoionization by solar Lyman alpha and x-rays, and the typical electron density is about 10^3 to 10^4 cm^{-3}.

E-region The E-layer extends from about 90 to 140 kilometers, and the peak daytime electron density is about 10^5 cm^{-3}.

F-region This region has two maxima in the daytime called the F_1 and the F_2 peaks, with maximum daytime electron densities of about 10^6 and 10^7 cm^{-3}, respectively, for solar maximum conditions. These are at least an order of magnitude less under solar minimum conditions. The F_2 layer virtually disappears at night. The F_1 layer is from about 140 to 200 kilometers, and the F_2 layer is from about 200 to 400 kilometers.

Topside ionosphere The region above the F_2 peak, where the electron density decreases slowly with altitude toward the exospheric level, may be referred to as the topside ionosphere. Eventually, in the magnetosphere, the atmosphere consists largely of atomic hydrogen ions and electrons.

6.2 Atmospheric models

The need for common sources of the properties of the atmosphere has lead to the development of standard atmospheres. The *U. S. Standard Atmosphere, 1976*, 1976, remains very useful, particularly up to the mesopause, and it

6.2. ATMOSPHERIC MODELS

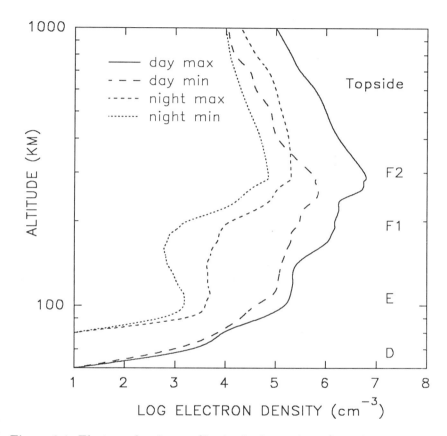

Figure 6.4: Electron density profiles in the ionosphere (based on data in the *Handbook of Geophysics and the Space Environment*)

has been included in transmission and radiance codes such as LOWTRAN 7. Unless mentioned, a standard atmosphere is for an idealized and steady-state situation at moderate solar activity.

In the thermosphere and much of the ionosphere, the MSIS model is used. This model is a large data base of empirical measurements classified by geographic location, geomagnetic activity, solar activity, solar zenith angle, and univeral time, *Hedin*, 1987. MSIS is an acronym for mass spectrometry and incoherent scatter. The results from these two techniques for measuring atmospheric properties were used as inputs to the model. Detailed tables resulting from the MSIS-86 model for days near the solar maximum and minimum and for a mid-cycle case are given by *Rees*, 1989. The MSIS approach has recently been extended to lower altitudes, *Hedin*, 1991.

A model for the thermosphere and exosphere based largely on satellite drag data and published by *Jacchia*, 1971, has been widely used in the past.

With the advent of greater computational capacity, global models including transport processes are being constructed for the thermosphere and ionosphere. These models will enable global effects to be followed and predicted, and they serve as a vehicle with which to organize our understanding of much of the atmosphere. It is difficult to predict the availability to many users of the computer facilities needed for such models, and for the next few years they will be available primarily to specialists. Work on such models has been published by *Fuller-Rowell and Rees*, 1980; *Dickinson et al.*, 1981, 1984; *Roble and Ridley*, 1987; and *Roble et al.*, 1987, 1988. The model in the last reference is called the thermospheric/ionospheric general circulation model, or the TIGCM model. An earlier version is called the thermospheric general circulation model, or TGCM. Publications continue to appear at a rapid pace, but further discussion is beyond our scope.

Global circulation models are important to us because some of the remote sensing applications discussed in Chapters 14 through 17 may be incorporated into these models as real time inputs. In this manner, we can expect that ultraviolet methods will play a role in the comprehensive understanding of the atmosphere and its changes.

6.3 References

Brasseur, G. and S. Solomon, *Aeronomy of the Middle Atmosphere*, Reidel, 1984.

6.3. REFERENCES

Champion, K. S. W., A. E. Cole, and A. J. Kantor, Standard and Reference Atmospheres, Chapter 14 in *Handbook of Geophysics and the Space Environment*, A.S. Jursa, Scientific Editor, Air Force Geophysics Laboratory, NTIS Document Accession Number ADA 167000, 1985.

Dickinson, R. E., E. C. Ridley, and R. G. Roble, A three-dimensional, time-dependent general circulation model of the thermosphere, *J. Geophys. Res., 86*, 1499–1512, 1981.

Dickinson, R. E., E. C. Ridley, and R. G. Roble, Thermospheric general circulation with coupled dynamics and composition, *J. Atmos. Sci., 41*, 205–219, 1984.

Fuller-Rowell, T. J. and D. Rees, A three-dimensional, time-dependent global model of the thermosphere, *J. Atmos. Sci., 37*, 2545–2657, 1980.

Hedin, A. E., MSIS-86 Thermospheric Model, *J. Geophys. Res., 92*, 4649–4662, 1987.

Hedin, A. E., Extension of the MSIS thermospheric model into the middle and lower atmosphere, *J. Geophys. Res., 96*, 1159–1172, 1991.

Jacchia, L. G., Revised static models of the thermosphere and exosphere with empirical temperature profiles, *Special Report 332, Smithsonian Astrophysical Obsrvatory*, Cambridge, Mass., 1971.

Rees, M. H., *Physics and Chemistry of the Upper Atmosphere*, Cambridge U. Press, 1989.

Roble, R. G. and E. C. Ridley, An auroral model for the NCAR thermospheric general circulation model (TGCM), *Annales. Geophysicae, 5A(6)*, 369–382, 1987.

Roble, R. G., E. C. Ridley, and R. E. Dickinson, On the global mean structure of the thermosphere, *J. Geophys. Res., 92*, 8745–8758, 1987.

Roble, R. G., E. C. Ridley, A. D. Richmond, and R. E. Dickinson, A coupled thermosphere/ionosphere general circulation model, *Geophys. Res. Lett., 15*, 1325–1328, 1988.

U. S. Standard Atmosphere, 1976, NOAA-S/T 76-1562, U. S. Government Printing Office, Washington, D.C., 1976.

Warneck, P., *Chemistry of the Natural Atmosphere*, Academic Press, 1988.

Chapter 7

Solar Photoabsorption

The ultraviolet sun is a variable star, and its emission in the ultraviolet is the subject of much research. The initial difficulty with ultraviolet solar flux measurements is that over most of the range it is necessary to make the measurements in the upper atmosphere, because of atmospheric absorption at lower altitudes. While there is nothing fundamentally difficult with these measurements, they are complicated by the need for access to space of reliable, calibrated sensors.

When these requirements are coupled with the large change in emission over the eleven year solar cycle, at least in the extreme ultraviolet, the difficulty of acquiring a reliable data base in the thirty or so years of measurements from space is apparent. However, flux value uncertainties have been steadily reduced, so that major differences among measurements in the future are unlikely.

Knowledge about the solar flux throughout the Earth's atmosphere is crucial to the development of remote sensing methods and transmission and radiance models. Therefore, this chapter introduces the absorption law, photodissociation, and photoionization. The solar flux is eventually completely absorbed over much of the ultraviolet. The sun is also used in remote sensing of the atmosphere as a light source for occultation measurements, pointing direction determination, and in-flight calibration.

The solar flux measurements given in this chapter are largely based on the work of H. E. Hinteregger, L. J. Heroux, L. A. Hall, D. Bedo, and co-workers at the Air Force Geophysics Laboratory, *Hall et al.*, 1985, as included in their chapter of the *Handbook of Geophysics and the Space Environment*. The handbook also has a useful introductory article on the sun, *Altrock et al.*, 1985. More complete discussions of the solar output in the UV and at other wavelengths are given in *White*, 1977. A recent review was given at the VUV Radiation Physics conference by *Rottman*, 1990.

7.1 Quiet sun flux values

For our purposes, the primary interest in the sun is the full disk irradiance arriving at the top of the Earth's atmosphere. An introduction to the physics of the sun itself is given by *Stix*, 1989.

The solar flux at wavelengths longer than the MUV (300 nm to 10 micrometers) is given in Figure 7.1. The flux, given in the irradiance units $milliwatt\, cm^{-2}\, micrometer^{-1}$, reaches a maximum between 400 and 500 nm. The data are from a summary by *Pierce and Allen*, 1977, of the measurements of Labs and Neckel between 300 nm and 3 μm and of Thekaekara at wavelengths longer than 3 μm. Although this radiation is not absorbed in the Earth's atmosphere, the flux at ground level is attenuated by scattering. Corrections for the air mass are needed for ground level flux values.

The solar flux is shown from 1 to 1500 Å in Figure 7.2 and from 1500 to 3000 Å in Figure 7.3. The flux is given as summed over 20 Å intervals. There is a marked change in appearance, even at this coarse resolution, at about 1500 Å. The solar spectrum at about this point changes from a continuum with complicated Franhofer absorption superimposed to a spectrum of discrete emission lines. Note the large change in flux across this range, dropping about five orders of magnitude from 3000 to 1500 Å. In the shorter wavelength range, the hydrogen Lyman-α line at 1215.67 Å and the He II resonance line at 303.78 Å dominate their regions. These flux values are largely based on measurements from the Atmospheric Explorer satellites, *Hinteregger et al.*, 1973; *Hinteregger*, 1976; *Heroux and Hinteregger*, 1978; *Hinteregger*, 1981; *Hinteregger, et al.*, 1981; *Hall et al.*, 1985.

The large scale variation in the solar flux with wavelength is shown well by Figures 7.1, 7.2, and 7.3, but for detailed work, spectra are needed with the best available resolution and accuracy. A solar reference spectrum from *Hinteregger et al.*, 1981, gives 810 wavelengths of solar lines between 18.62 and 1050.01 Å in the EUV. This compilation is known as solar reference spectrum SC # 21REFW, and it is available from the National Space Science Data Center, NOAA, Boulder, Colorado. Additional solar reference spectra are available from the National Space Science Data Center, NASA, Goddard SFC, *Bilitza*, 1990.

7.1. QUIET SUN FLUX VALUES

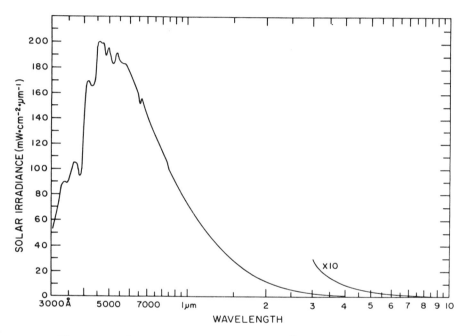

Figure 7.1: Solar flux longer than 3000 Å(*Hall, et al.*, 1985)

CHAPTER 7. SOLAR PHOTOABSORPTION

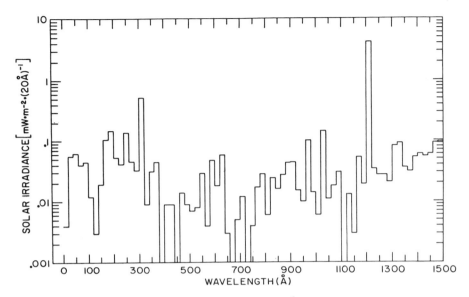

Figure 7.2: Solar flux from 1 to 1500 Å (*Hall et al.*, 1985)

A widely used set of solar flux values has been published by *VanHoosier et al.*, 1988. These solar flux values have been used over the 2000 to 3500 Å region in the LOWTRAN 7 model *Anderson et al.*, 1989, which is discussed further in Chapter 13. The flux values are at 1.5 Å resolution and are included in the model down to a wavelength of 1740 Å. The measurements of VanHoosier *et al.*, overlap the Hinteregger spectrum mentioned in the last paragraph. In the region from 3500 to 8600 Å, the measurements of *Neckel and Labs*, 1984, were used for LOWTRAN 7.

The review by *Rottman*, 1990, provides an excellent assessment of the current best measurements. References are also given to this author's experiment on the Solar Mesosphere Explorer, which has been important in determining MUV fluxes and variability. Recently, balloon measurements have been extrapolated to above the atmosphere to provide a solar spectrum between 200 and 310 nm at 0.01 nm resolution, *Hall and Anderson*, 1991. New measurements have also recently been reported by *Schmidtke et al.*, 1991, from the San Marco-5 satellite.

7.1. QUIET SUN FLUX VALUES

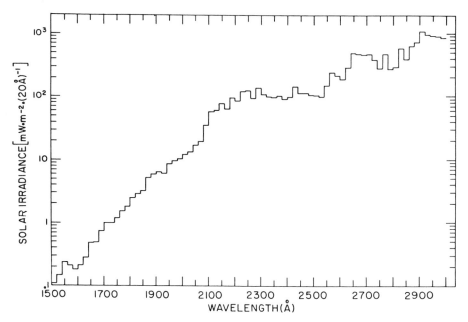

Figure 7.3: Solar flux from 1500 to 3000 Å(*Hall et al.*, 1985)

7.2 Solar flux variability

The solar output varies in many regions of the electromagnetic spectrum, and it is remarkable that the visible sun has so little variability. The concept of a solar constant giving the solar energy input at the Earth from the sun is valid, at least for the 99% of the solar energy at wavelengths longer than 300 nm. The solar constant has been given as $1373 \pm 20\, watt/meter^2$ by *Frohlich*, 1977.

This brief discussion is an introduction for nonspecialists to the subject of solar UV variability. The variable part of the solar ultraviolet is primarily in the EUV and FUV. In terms of energy, this radiation is insignificant compared to the energy input to the Earth from other wavelength regions. However, the variable EUV and FUV radiation is extremely important in terms of its effects in the thermosphere and ionosphere.

Since the relative variability in the near and middle ultraviolet over the solar cycle is less than for the far and extreme ultraviolet, Figures 7.1 and 7.2 should be useful as a first approximation throughout the solar cycle. These are not at the highest resolution available, however. At higher resolution, the Franhofer absorption structure in the solar spectrum may have to be considered in detail. In the far and especially in the extreme ultraviolet, there can be large changes over the solar cycle. There is also a monthly variation in EUV flux, due to the solar rotation, and possibly other, weaker cyclic variability.

There has been a long interest in using ground observables, such as the 10.7 cm radiowave solar emission and the sunspot number, as "proxies" for space measurements of the solar UV flux. Past ground observables records are available, and development of methods to correlate these records with the solar UV flux would allow extrapolation of the solar intensity to previous years and even centuries.

Unfortunately, the correlation does not seem to be simple. As an introduction to these model studies, the reader is referred to *Donnelly et al.*, 1983, 1985, and 1986. These references also provide valuable summations of solar flux measurements that may not be readily available elsewhere.

Solar extreme ultraviolet flux models of the variability based on measurements in space have been given by *Torr et al.*, 1979, 1980; *Hinteregger et al.*, 1981; *Schmidtke*, 1984; and *Tobiska and Barth*, 1990. A review section of several papers covering solar variability and its atmospheric effects is recommended, *Lean*, 1987. Solar variability itself has recently been reviewed, *Lean*, 1991.

7.3 Extinction and absorption

The extinction of ultraviolet radiation as it passes through the atmosphere can be from absorption or from scattering. The incremental decrease in a beam of photons passing through an incremental path in the medium is as follows:

$$dI(\lambda) = -k(\lambda)I(\lambda)dl, \tag{7.1}$$

where $I(\lambda)$ represents a beam of radiation in *photons* sec^{-1} cm^{-2}; $k(\lambda)$ is the extinction coefficient in units of reciprocal length such as cm^{-1}; and l is the path length in cm.

As we shall see later, much of the complexity in atmospheric extinction problems results from situations where Eq. 7.1 appears not to be followed. These situations may arise as a result of wavelength resolution insufficient for the problem or from changes in composition, such as the dimerization that can sometimes occur at higher densities.

We will assume for the remainder of this chapter that the extinction of the beam is only due to photoabsorption, unless specifically stated otherwise, and that scattering is insignificant. The Rayleigh scattering cross section is about 2×10^{-25} cm^2 at 300 nm, and it rises to about 3×10^{-22} cm^2 at 50 nm. These values are usually several orders of magnitude below the absorption cross sections for the major gases over most of the ultraviolet. Scattering is discussed further in Chapter 11.

Since absorption by gaseous constituents only is being considered here, the extinction coefficent $k(\lambda)$ in the above equation can be replaced as follows:

$$k(\lambda) = \sigma(\lambda)n. \tag{7.2}$$

In this case, $\sigma(\lambda)$ is the **absorption cross section** with units of cm^2, and n is the local number density of absorbers in units of $absorbers/cm^3$. The absorption cross section, σ, and other cross sections for photoionization, etc., used in atmospheric problems are usually given in units of cm^2. Another cross section unit sometimes encountered is the Megabarn, Mb, which is $10^{-18} cm^2$.

Substituting the equation for the absorption cross section into Eq. 7.1 and integrating along a uniform column of gas, the following equation results:

$$I(\lambda) = I_o(\lambda)e^{-\sigma(\lambda)nl}. \tag{7.3}$$

In this equation, $I(\lambda)$ is the flux transmitted through the column, $I_o(\lambda)$ is the flux incident on the column, σ is the absorption cross section in cm^2,

n is the number density of the absorber in cm^{-3}, and l is the length of the absorption path in cm.

The names of Bouguer, Lambert, and Beer are associated with the exponential extinction relationships of Eqs. 7.1 and 7.3. An interesting short account of the contributions of each of these individuals has been given by *Perrin*, 1948.

The *absorption coefficient*, $K(\lambda)$, as usually defined, is equal to $\sigma(\lambda) n_o$, and it has units of cm^{-1}. In this case, the exponential in Eq. 7.3, $\sigma(\lambda) nl$, is replaced by $K(\lambda) l_s$, where l_s is the path length reduced to standard temperature and pressure, or STP, (0 °C, 1 atmosphere). The quantity n_o, Loschmidt's number, is $2.687 \times 10^{19}\ molecules/cm^3$, which is at STP.

In Eq. 7.3, it may be convenient to use N, the column density of the absorber in units of $absorbers/cm^2$. For an absorbing medium of uniform density, $N = nl$. Substituting into Eq. 7.3,

$$I(\lambda) = I_o(\lambda) e^{-\sigma N}. \tag{7.4}$$

For the case where the density is nonuniform, N must be integrated over the altitude range and along the slant path of interest. Therefore, in the atmosphere,

$$N = \int_{l_1}^{l_2} n(h)\, dl, \tag{7.5}$$

where $n(h)$ is the number density at the altitude h.

For most atmospheric transmission problems, including those involving solar absorption, the absorption is along a slant path. In the solar case, the *solar zenith angle*, χ, defines the direction of this slant path, and the equations given above must be modified. The solar zenith angle χ is the angle between the zenith or directly overhead direction and the sun. The relationship between an incremental absorption path length as measured along the zenith direction by h and the slant path in the solar direction as measured by l is $dl = \sec \chi\, dh$. For χ greater than about 70°, the fact that the Earth is spherical cannot be ignored and the Chapman function must be used. A valuable approximation to this function has been given by *Swider and Gardner*, 1967. It is more readily available on p. 108 of *Brasseur and Solomon*, 1984.

For atmospheric transmission modeling, the absorption equation must also be expanded to include all the absorbing constituents and then integrated across the wavelength region of interest. It is necessary to know the

7.3. EXTINCTION AND ABSORPTION

intensity of the source radiation, which is frequently solar radiation; the atmospheric density as a function of altitude; and the total absorption cross sections as a function of wavelength for all of the significant absorbers.

Combining the above equations, the transmitted solar flux $I_{\Delta\lambda,h}$ over the wavelength interval $\Delta\lambda = \lambda_2 - \lambda_1$ and at altitude h is

$$I_{\Delta\lambda,h} = \int_{\lambda_1}^{\lambda_2} I_o(\lambda) \sum_i \exp\left(-\sigma_i(\lambda) \int_h^\infty n_i(h) \sec\chi \, dh\right) d\lambda, \quad (7.6)$$

where $\sigma_i(\lambda)$ is the absorption cross section of the atmospheric constituent i and $n_i(h)$ is its number density at altitude h.

The reader should be aware of other common names for quantities associated with atmospheric extinction. The **transmission**, T, is frequently used. It is defined as $I(\lambda)/I_o(\lambda)$. The exponential in the absorption equation may be defined as τ, which is sometimes called the **optical depth**. This quantity is a convenient measure of the attenuation of the radiation along the absorption path.

$$\tau = \sigma(\lambda) N \quad (7.7)$$

An optical depth of unity is an indication of significant atmospheric absorption. For $\tau = 1$, $I_o(\lambda)$ has been reduced to $(1/e)I_o(\lambda)$.

Although its use is not encouraged here, the optical density, using the base 10 logarithm scale, is sometimes encountered in older literature, but its use is not recommended at this time. It is defined as follows:

$$\text{Optical Density (OD)} = \log\frac{I_o}{I} = 0.4343(\sigma N). \quad (7.8)$$

In each specific case, it is advisable to find the form of Eqs. 7.1 and 7.3 being used, as the same name may be used for different quantities.

The absorption of photons along a column results in the disappearance of the photon and the reappearance of the energy of the photon in other forms. The absorber may be excited to a higher energy state, dissociate into smaller neutral species, ionize to form a positively charged ion and an electron, or undergo some combination of these changes. The energy of the photon is thereby deposited in the atmosphere. The amount of energy deposited or the number of processes occurring depends on the number of photons absorbed along the path. The number of photons absorbed along the path is found from Eq. 7.3 as follows:

$$I_o(\lambda) - I(\lambda) = I_o(\lambda)[1 - exp(-\sigma(\lambda)N)]. \quad (7.9)$$

For the case of weak absorption, the above expression becomes:

$$I_o(\lambda) - I(\lambda) = I_o(\lambda)\sigma(\lambda)N. \tag{7.10}$$

This version is also useful to find local absorption effects. In all cases it must be understood that it is necessary to include all significant atmospheric absorbers and to integrate over the wavelength range of interest.

Solar absorption occurs in the atmosphere in layers, to a first approximation, as first pointed out by *Chapman*, 1931. The exponential decrease in density as a function of altitude combined with the exponential nature of the absorption of incoming photons combine to produce a reasonably sharp maximum in absorption at a certain altitude. For larger solar zenith angles away from the overhead sun position, the maximum is less pronounced and occurs at a higher altitude. The Chapman layer concept explains the formation of layers of ionospheric electron density, thermospheric atomic oxygen, and stratospheric ozone, although transport and collision processes will modify the distribution both with altitude and with global location. Further discussion of Chapman layers and the accurate calculation of solar zenith angle effects are found in books on the atmosphere such as *Brasseur and Solomon*, 1984.

Equation 7.3 may appear to give an inaccurate value for the transmission for a number of reasons. A known molecular composition of absorber is assumed to be present. However, there may be changes in the composition due to the formation of weakly bound cluster molecules at certain temperatures and pressures. In addition, a significant population in another vibrational or rotational state may be a factor at some atmospheric temperatures.

Another difficulty with application of Eq. 7.3 concerns **wavelength resolution**. There is a bandwidth associated with every absorption problem, resulting from the characteristics of the source, the sensing instrument, or a value chosen for the model. The simple Eq. 7.3 given above assumes that the cross section does not vary significantly across this bandwidth. When the bandwidth is too large to make this assumption accurate, the transmission attenuation of a beam will not be given by Eq. 7.3. It should be pointed out that the difficulty is with our knowledge of the source and absorber and not with the physical principles. If the bandwidth is inadequate, improved measurements or a more detailed theoretical calculation is ultimately required.

Current problems with inadequate bandwidth information for atmospheric species are usually confined to narrow lines found in the rotational structure of molecular bands or in atomic multiplets. Modeling of solar emission line attenuation in the extreme ultraviolet may be difficult due to a solar line or multiplet of narrow width and poorly known shape being absorbed in a region of molecular bands with absorption cross section shapes distorted by perturbations and autoionization.

When the bandwidth is inadequate, the exponential nature of Eq. 7.3 makes it impossible to measure or use a simple average cross section that will accurately describe the transmission. The actual transmission as a function of column density measured when the resolution is inadequate is sometimes used instead of the cross section. This is sometimes called a **curve-of-growth** approach. The effect of an inadequate bandwidth is also sometimes called **radiation hardening**, because the part of the beam with the higher cross section is absorbed first along the transmission path. However, a portion of the beam appears to be harder, with a smaller cross section. Thus, a portion of the beam is found to be transmitted to altitudes lower than predicted with some sort of average cross section.

Finally, it must be pointed out that for many problems in the troposphere and at the surface of the Earth, multiple scattering by both the molecules in the atmosphere and also by the ever present aerosols and other particles must be considered. These effects become very important for laser beam transmission, for example.

7.4 Photodissociation and photoionization

The absorption cross section, σ, as defined in this book, is the sum of all the separate cross sections for processes that are possible resulting from a single photon absorption event at the wavelength of the absorbed photon. This sum is frequently called the total absorption cross section to emphasize the point. The two most important of these detailed processes, and the ones that lead to the most significant changes in the upper atmosphere compared to the surface atmosphere, are **photodissociation** and **photoionization**. These processes result in the formation of new species, while the excitation process, which may occur at the same wavelength as well, results in a change in the electronic or other energy level of the constituent.

Absorption leading to **excitation** is indicated in the spectrum by absorption lines, possibly with multiplet structure, for atoms, and by electronic bands consisting of many lines due to vibrational and rotational energy levels, for molecules. Since the energy of the photon must match the energy difference between the lower and the upper energy level, the absorption occurs as discrete lines or bands in a narrow range of wavelengths. The usual result of excitation is the emission of a photon at the same wavelength and a return of the absorber to its original energy level.

Photodissociation and photoionization are indicated in a spectrum by continuous absorption over an extended range of wavelengths. In contrast to excitation, which is indicated by absorption at discrete wavelengths, the absorption due to photodissociation and photoionization occurs in an absorption continuum. For a continuum, the transition is not between two well-defined energy levels but between a specific lower energy level and any energy greater than the threshold energy for a given process. An exact match in energy between two levels is not needed, since excess energy above the threshold can be accommodated by an increase in the kinetic energies of the products. In practice, many different processes may occur together in the same wavelength region, with discrete absorption in lines or bands overlapping one or more continua.

Photodissociation results in the severing of chemical bonds and the production of neutral dissociation products, as follows:

$$XY + h\nu \longrightarrow X + Y. \tag{7.11}$$

Both X and Y can be atomic or molecular; they may be in excited states (electronic, vibrational, rotational); and they can have excess kinetic energy. The products in excited electronic states are usually much more reactive than the ground state, making them more significant in the atmosphere. The fraction of the absorption processes that results in the formation of a product in a specific electronic state X is called the quantum yield, $Q_X(\lambda)$. The cross section for this process is

$$\sigma_X(\lambda) = Q_X(\lambda)\sigma(\lambda), \tag{7.12}$$

where $Q_X(\lambda)$ is the quantum yield, or simply the yield, for the production of X at a given wavelength, and $\sigma(\lambda)$ is the total absorption cross section. A product excited to a state above the ground state is sometimes indicated by some kind of superscript, such as X^*.

Photodissociation is possible at all wavelengths less than the threshold corresponding to the least energetic photodissociation process for the molecule in question, although whether the absorption will be significant depends on the cross section. Since the absorption occurs over a continuous

7.4. PHOTODISSOCIATION AND PHOTOIONIZATION

range of wavelengths, rather than at discrete wavelengths as in the case of excitation, photodissociation leads to continua in the cross section curve. The difference in energy between that possessed by the absorbed photon and the energy levels of the products is taken up in the kinetic energy of the products.

Photoionization involves the ejection of an electron from the absorbing atom or molecule as follows:

$$XY + h\nu \longrightarrow XY^+ + e. \qquad (7.13)$$

Dissociative photoionization, where one product of a dissociation process is ionized and the other remains neutral, happens as well.

The lowest energy required for photoionization to occur in a molecule is higher than the lowest energy photodissociation process, at least for the atmospheric molecules. Photoionization therefore is more important at shorter wavelengths and higher altitudes than photodissociation. The part of the total absorption cross section that leads to ionization is called the photoionization cross section, $\sigma_I(\lambda)$. This part can be expressed as a fraction called the photoionization yield, or simply ionization yield, $y(\lambda)$, and it is related to the total absorption cross section as follows:

$$\sigma_I(\lambda) = y(\lambda)\sigma(\lambda). \qquad (7.14)$$

As in the photodissociation case, there may be a number of electronic states of the ion involved, and there are specific yields for each wavelength where the process can occur that give the amounts of these products produced. The excess kinetic energy is taken up primarily by the electron in photoionization.

There are usually several possible electronic states for both neutral and ionized products of photoabsorption, and each can be associated with a quantum yield or ionization yield. The yield can also be called the **branching ratio**. This is the fraction of absorption events that leads to a specific product. It is most useful in the extreme ultraviolet, where the energy carried by the photon allows many ionized and neutral states of the products to be reached. The term branching ratio is also used to describe the distribution of emission from a given upper energy level to several lower energy levels.

Two important types of photon interaction that contribute to the complexity of absorption at ultraviolet wavelengths are **predissociation** and **autoionization**. In these cases, photon absorption places the absorber in a discrete energy level that lies in energy higher than the dissociation or ionization threshold. There is thus a continuous energy range available for a radiationless transition from the initial discrete energy level to the continuum. This transition can be very much faster than an allowed transition to a lower discrete level. The excited absorber thus undergoes dissociation or ionization, resulting in a broadened and possibly perturbed discrete absorption structure in addition to the continuum at this wavelength. Autoionization in molecules is sometimes called preionization.

7.5 References

Altrock, R. C., H. L. DeMastus, J. W. Evans, S. L. Keil, D. F. Neidig, R. R. Radick, and G. W. Simon, The sun, Chapter 1, *Handbook of Geophysics and the Space Environment*, A. S. Jursa, Editor, Air Force Geophysics Laboratory, NTIS Document Accession Number ADA 167000, 1985.

Anderson, G. P., F. X. Kneizys, E. P. Shettle, L. W. Abreu, J. H. Chetwynd, R. E. Huffman, and L. A. Hall, LOWTRAN 7 spectral simulations in the UV, *Applied Opt. (in preparation)*; see also *ibid., Conference Proceedings No. 254, Atmospheric propagation in the UV, visible, IR, and MM-wave region and related systems aspects, AGARD, NATO*, 25-1 to 25-9, 1989, (Available from the National Technical Information Service, (NTIS), 5285 Port Royal Road, Springfield, VA 22161).

Bilitza, B., *Solar-Terrestrial Models and Applications Software*, NSSDC/ WDC A R&S 90-19, National Space Science Data Center, Goddard SFC, 1990.

Brasseur, G. and S. Solomon, *Aeronomy of the Middle Atmosphere*, Reidel, 1984.

Chapman, S., The absorption and dissociative or ionizing effect of monochromatic radiations in an atmosphere on a rotating Earth, *Proc. Phys. Soc., 43*, 483, 1931.

Donnelly, R. F., D. F. Heath, J. L. Lean, and G. J. Rottman, Differences in the temporal variations of solar UV flux, 10.7 cm solar radio flux, sunspot number and Ca K plage data caused by solar rotation and active region evolution, *J. Geophys. Res., 88*, 9883–9888, 1983.

7.5. REFERENCES

Donnelly, R. F., J. W. Harvey, D. F. Heath, and T. P. Repoff, Temporal characteristics of the solar UV flux and He I line at 1083 nm, *J. Geophys. Res., 90*, 6267–6273, 1985.

Donnelly,, R. F., H. E. Hinteregger, and D. F. Heath, Temporal variations of solar EUV, UV, and 10,830 Å radiations, *J. Geophys. Res., 91*, 5567–5578, 1986.

Frohlich, C., Contemporary measures of the solar constant, in *The Solar Output and Its Variation*, O. R. White, editor, Colorado Associated University Press, 1977.

Hall, L. A. and G. P. Anderson, High resolution solar spectrum between 2000 and 3100 Ångstroms, *J. Geophys. Res., 96*, 12,927–12,931, 1991.

Hall, L. A., L. J. Heroux, and H. E. Hinteregger, Solar ultraviolet irradiance, Chapter 2, *Handbook of Geophysics and the Space Environment*, A. S. Jursa, Editor, Air Force Geophysics Laboratory, NTIS Document Accession Number ADA 167000, 1985.

Heroux, L. and H. E. Hinteregger, Aeronomical reference spectrum for solar UV below 2000 Å, *J. Geophys. Res., 83*, 5305–5308, 1978.

Hinteregger, H. E., EUV fluxes in the solar spectrum below 2000 Å, *J. Atm. Terr. Phys., 38*, 791–806, 1976.

Hinteregger, H. E., Representations of solar EUV flux for aeronomical applications, *Adv. Spa. Res., 1*, 39, 1981.

Hinteregger, H. E., D. E. Bedo, and J. E. Manson, The EUV spectrophotometer on Atmospheric Explorer, *Radio Sci., 8*, 349, 1973.

Hinteregger, H. E., K. Fukui, and B. R. Gilson, Observational, reference and model data on solar EUV, from measurements on AE-E, *Geophys. Res. Lett., 8*, 1147, 1981.

Lean, J. L., Solar ultraviolet irradiance variations: A review, *J. Geophys. Res., 92*, 839–868, 1987.

Lean, J. L., Variations in the sun's radiative output, *Rev. Geophys., 29*, 505–535, 1991.

Neckel, H. and D. Labs, The solar radiation between 3300 and 12500 Å, *Solar Physics, 90*, 205–258, 1984.

Perrin, F. H., Whose absorption law? *J. Opt. Soc. Am.*, *38*, 72–74, 1948.

Pierce, A. K. and R. G. Allen, The solar spectrum between 0.3 and 10 μm, in *The Solar Output and Its Variation*, O. R. White, Editor, Colorado Associated University Press, 1977.

Rottman, G. J., Recent advances in VUV observations of the sun, *Physica Scripta*, *T31*, 199–207, 1990.

Schmidtke, G., Modelling of the solar extreme ultraviolet irradiance for aeronomic applications, *Handb. Phys.*, *XLIX(7)*, *Geophysics III, Part VII*, 1, 1984.

Schmidtke, G., H. Doll, C. Wita, and S. Chakrabarti, Solar EUV/UV and equatorial airglow measurements from San Marco-5, *J. Atm. Terr. Phys.*, *53*, 781–785, 1991.

Stix, M., *The Sun, An Introduction*, Springer-Verlag, 1989.

Swider, W. and M. E. Gardner, On the accuracy of certain approximations for the Chapman function, *Environmental Research Papers No. 272*, Air Force Cambridge Research Laboratories, Bedford, Mass., 1967.

Tobiska, W. K. and C. A. Barth, A solar EUV flux model, *J. Geophys. Res.*, *95*, 8243–8251, 1990.

Torr, M. R., D. G. Torr, and H. E. Hinteregger, Solar flux variability in the Schumann-Runge continuum as a function of solar cycle 21, *J. Geophys. Res.*, *85*, 6063, 1980.

Torr, M. R., D. G. Torr, R. A. Ong, and H. E. Hinteregger, Ionization frequencies for major thermospheric constituents as a function of solar cycle 21, *Geophys. Res. Lett.*, *6*, 771, 1979.

VanHoosier, M. E., J. D. Bartoe, G. E. Brueckner, and D. K. Prinz, Absolute solar spectral irradiance monitor—SUSIM— Experiment on board Spacelab 2, *Astro. Lett. and Communications*, *27*, 163–168, 1988.

White, O. R., Editor, *The Solar Output and Its Variation*, Colorado Associated University Press, 1977.

Chapter 8

Photon Cross Sections

Photon cross sections are required for a quantitative understanding of the atmosphere. Photoabsorption determines the altitude region where solar radiation is deposited and the transmission of ultraviolet in the atmosphere. In addition, occultation and similar remote sensing methods use photoabsorption to measure atmospheric concentrations. Photoionization cross sections are needed for use at ionospheric altitudes.

This chapter is a brief review of important atmospheric photoabsorption and photoionization cross sections of the major constituents: ozone, molecular oxygen, molecular nitrogen, and atomic oxygen. The goal is to provide enough information to understand remote sensing methods described later in this book.

Following solar absorption, chemical reactions occur among the products and the ambient atmosphere, leading ultimately to the atmosphere as we find it. These neutral, ionic, and electron reactions are outside our scope, but are discussed in aeronomy texts.

Atmospheric photoabsorption between roughly 10 and 100 km is covered by *Brasseur and Solomon*, 1984 and by *Warneck*, 1988. The problems at higher altitudes are considerably different, as described by *Rees*, 1989, for the region upward from about 90 km, and by *Banks and Kockarts*, 1973.

Much of this chapter is based on the results of laboratory research. For laboratory measurements of use in aeronomy, see *Schiff*, 1969, for an older but still useful introduction. Spectroscopic nomenclature is explained in references such as *Herzberg*, 1950; and *Huber and Herzberg*, 1979. Photon processes are discussed by *Berkowitz*, 1979. The wide range of photon measurements now possible with lasers, synchrotron sources, and beam techniques has been described recently by *McDaniel*, 1989.

8.1 Energy levels and equivalent wavelengths

The energy levels and associated equivalent wavelengths of atmospheric constituents provide the framework for understanding photon interactions and chemical reactions in the atmosphere.

In Table 8.1, the most frequently needed energy levels and threshold wavelengths are given for the major photodissociation and photoionization processes. These are the minimum energies required before the process shown can occur. However, the cross section, the number density relative to other absorbers, and the availability of significant photon flux at the altitude in question will determine the importance of any specific process at any specific place in the atmosphere.

The results of the absorption of a photon by an atmospheric gas depends on the energy of the photon relative to the energy levels available to the absorber and products. A process such as

$$O_2 + h\nu \longrightarrow O + O$$

has an associated total absorption cross section, σ, as discussed in the last chapter. However, for most atmospheric problems other than transmission, a total energy state definition of the O_2 reactant and the O products is needed. A complete description will also give the kinetic energies and angular distributions of the reactants and products. For initial considerations and for the cross sections given here, the absorber is in the ground state at room temperature. Products are discussed in many cases, where they are known.

8.2 Absorption and ionization cross sections

The total absorption cross sections for the major constituents ozone (O_3), molecular oxygen (O_2), molecular nitrogen (N_2), and atomic oxygen (O) are given in Figures 8.1 and 8.2, *Huffman*, 1985. These figures show the overall or total absorption cross section in cm^2 in the regions of the ultraviolet of primary importance in the atmosphere. See Chapter 7 for a discussion of cross sections and the absorption law.

The cross sections shown in Figures 8.1 and 8.2 are intended to be introductory in nature in order to show where the dominant absorption occurs

8.2. ABSORPTION AND IONIZATION CROSS SECTIONS

Table 8.1: Important atmospheric energy thresholds (from ground states)

Transition	λ (nm)	E (eV)	Comment
$O_3\,(\tilde{X}\,^1A_1) \to O_2\,(X\,^3\Sigma_g^-) + O\,(^3P)$	1192.	1.04	Dissociation
$O_3\,(\tilde{X}\,^1A_1) \to O_2\,(a\,^1\Delta) + O\,(^1D)$	311.	3.99	Hartley cont.
$O_2\,(X\,^3\Sigma_g^-) \to O\,(^3P) + O\,(^3P)$	242.4	5.115	Herzberg cont.
$O_2\,(X\,^3\Sigma_g^-) \to O\,(^1D) + O\,(^3P)$	175.1	7.081	S-R cont.
$O_2\,(X\,^3\Sigma_g^-) \to O_2^+\,(X\,^2\Pi_g) + e$	102.8	12.06	Photoion. cont.
$N_2\,(X\,^1\Sigma_g^+) \to N\,(^4S) + N\,(^4S)$	127.0	9.759	Dissociation
$N_2\,(X\,^1\Sigma_g^+) \to N_2^+\,(X\,^2\Sigma_g^+) + e$	79.58	15.580	Photoion. cont.
$O\,(^3P) \to O^+\,(^4S) + e$	91.04	13.618	Photoion. cont.
$NO\,(X\,^2\Pi) \to NO^+\,(X) + e$	133.8	9.267	Photoion. cont.

for each major species. As such, they are useful in gaining an overall quantitative understanding of how strongly the constituents absorb photons on a per molecule basis and for making first approximations. These large scale curves cannot show details of band and line structure. Regions where the absorption is dominated by discrete as opposed to continuous absorption are indicated with vertical lines. In these regions, more details concerning the absorption cross sections must be sought through the references.

There is no better introduction to photon absorption in the upper atmosphere at wavelengths shorter than 200 nm than the review by *Watanabe*, 1958, although there are now improved measurements in most cases. Cross section reviews for atmospheric absorption applications have been given by *Schoen*, 1969; *Huffman*, 1969; *Hudson*, 1971; and *Stolarski and Johnson*, 1972. A widely used compilation for the wavelength region from the molecular oxygen photoionization threshold to higher energies is given by *Kirby-Docken et al.*, 1979, which covers the 3.4 to 102.7 nm region and gives branching ratios of the products. Reviews and averaged values found useful in many studies are given by *Torr et al.*, 1979, and *Torr and Torr*, 1985. A valuable recent review for use in calculations of photoelectron production in the 1.9 to 105 nm range has been given by *Conway*, 1988.

OZONE

Ozone absorption in the stratosphere shields the Earth's surface from solar near-ultraviolet wavelengths. Ozone absorption cross-sections important in the atmosphere are shown in Figure 8.1. The absorption begins weakly at about 360 nm and reaches a maximum in the 250 nm region. At wavelengths less than about 200 nm, ozone absorption is usually unimportant in the atmosphere compared to the dominant molecular oxygen absorption. Stratospheric ozone prevents all solar radiation at wavelengths less than about 280 nm from reaching the surface of the Earth.

The ozone absorption beginning near 360 nm is a group of diffuse bands called the Huggins bands. This absorption dissociates ozone into molecular oxygen and atomic oxygen, with both products in their ground electronic states.

$$O_3\,(\tilde{X}\,^1A_1) + h\nu \longrightarrow O_2\,(X\,^3\Sigma_g^-) + O\,(2p^4\,^3P_{2,1,0})$$

The absorption cross sections in the bands are temperature dependent, and the bands are superimposed on weak continuous absorption. The Huggins bands are not shown separately in Figure 8.1, but they are available in *Anderson et al.*, 1989, and references therein.

The Hartley band and continuum region begins at about 310 nm, when it becomes energetically possible to produce an oxygen atom in the electronically excited $^1D_2$ state and the associated molecular oxygen in the first excited singlet state $a\,^1\Delta_g$. The atmospheric importance of the $^1D_2$ state is that it has sufficient energy to undergo an exothermic reaction with ground state molecular oxygen leading to the regeneration of ozone. The Hartley region cross sections do not significantly change in value over the range of temperatures found in the atmosphere, except near the long wavelength end. The cross section curve given in Figure 8.1 is essentially the same as measured by *Inn and Tanaka*, 1953. The curve has been modified only slightly by more recent measurements. Ozone also absorbs in the visible in the Chappius bands (440–770 nm) and in the infrared at 9.6 μm. These absorption features are also used for remote sensing, *Griggs*, 1966.

The importance of the ozone problem in the stratosphere has led to an extraordinary amount of concern about all the absorption processeses and reaction rates associated with it. The reader is referred to *DeMore et al.*, 1990, for a current review.

8.2. ABSORPTION AND IONIZATION CROSS SECTIONS

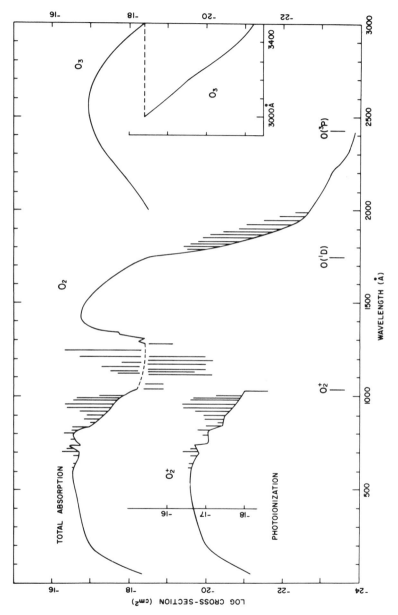

Figure 8.1: Absorption cross sections for molecular oxygen and ozone

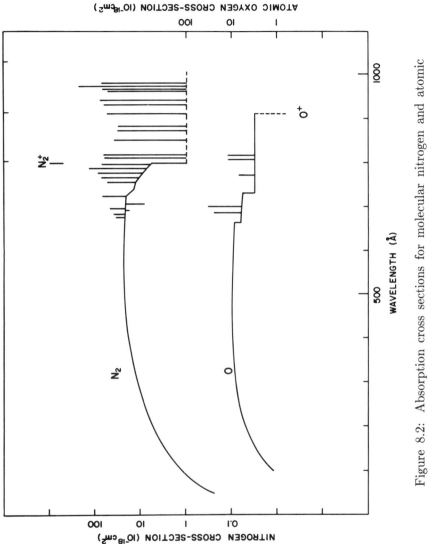

Figure 8.2: Absorption cross sections for molecular nitrogen and atomic oxygen

8.2. ABSORPTION AND IONIZATION CROSS SECTIONS

MOLECULAR OXYGEN

Molecular oxygen (O_2) absorption and ionization cross sections are summarized in Figure 8.1, and the principal products and energy relationships are shown in Table 8.1. These absorption cross sections, covering over six orders of magnitude, will be discussed proceeding from longer to shorter wavelengths.

There has been a great amount of work on oxygen absorption in the Herzberg and Schumann-Runge band regions, because of the central importance of the absorption of solar radiation by oxygen in determining the ozone concentrations. This massive amount of work at several laboratories cannot be reviewed here, but it can be traced through the comprehensive paper of *Nicolet et al.* 1989.

The molecular oxygen **HERZBERG CONTINUUM AND BANDS** are important in atmospheric absorption, even though the cross section is small relative to ozone cross sections in the Hartley band region, because of the larger number density of oxygen compared to ozone. As shown in Table 8.1, the products of the photoabsorption are two oxygen atoms in their ground states ^3P. The Herzberg continuum cross section is difficult to measure in the laboratory because the relatively small cross section requires long path lengths to obtain measurable attenuations. Another difficulty is the formation of weakly bound, absorbing dimers between an oxygen molecule and either another oxygen molecule or a nitrogen molecule. A general discussion of absorption in this region and choices made for the LOWTRAN 7 code is given by *Anderson et al.*, 1989.

The associated Herzberg bands adjoining the continuum on the longer wavelength side have recently been shown to be important in regions of the atmosphere where ozone absorption can be relatively weak, such as near the surface, *Trakhovsky et al.*, 1989a,b. These weak bands are not included in Figure 8.1.

The molecular oxygen **SCHUMANN-RUNGE BANDS** can be considered to begin at the (1,0) band, which is a relatively weak absorption feature at about 205 nm. The higher vibrational states of the band system dominate the absorption cross section in the region down to the onset of the continuum at about 176 nm. The most prominent bands are shown schematically as vertical lines in Figure 8.1.

The Schumann-Runge bands control the absorption in the mesosphere from about 65 to 90 km. They have represented a formidable challenge to measurement and modeling, because of the complexity of the rotational line structure of the bands. Measurements are now available, including the temperature dependence of the cross section and the absorption by the isotope $^{18}O^{16}O$.

Experimental work on the bands, which has been extensive recently, can be found through *Yoshino et al.*, 1989; *Lewis et al.*, 1986; and *Blake*, 1979. A comprehensive review and listing of the large amount of information now available is given by *Nicolet et al.*, 1989. This paper also provides detailed cross section curves utilizing the available experimental and theoretical investigations. It is likely that the most use of the large amount of information now available will be through UV transmission codes (see Chapter 13).

The **SCHUMANN-RUNGE CONTINUUM** extends from about 175 to 130 nm. It becomes more important than the Schumann-Runge band absorption at about 90 kilometers and above. This strong absorption is largely responsible for the conversion of molecular oxygen to the atomic form in the thermosphere. The absorption cross section reaches greater than $10^{-17} cm^2$, which is about the same as that reached in the ionization continuum (see Figure 8.1). Absorption in the continuum yields oxygen atoms in excited electronic states. The onset of the continuum coincides with the energy required to produce a ground state atom $O\ ^3P$ and an atom in the first electronic excited state $O\ ^1D$.

The Schumann-Runge continuum cross sections reported by *Watanabe*, 1958, have not changed significantly, although there have been a number of measurements since then. The temperature dependence has been measured by *Hudson et al.*, 1966, and by *Lean and Blake*, 1981.

The **ATMOSPHERIC WINDOW REGION** in the molecular oxygen absorption spectrum is from about 130 nm to the ionization threshold at about 102 nm. It consists of many strong, diffuse absorption bands coupled with regions between the bands of low absorption. The regions of low absorption are called atmospheric windows, because high energy solar UV can penetrate to lower altitudes than would be expected from averaged cross sections at these wavelengths. Figure 8.1 indicates the locations and strengths of many of these strong bands and deep windows.

The most important atmospheric effect associated with the windows is due to the location of a window at the strong hydrogen Lyman alpha solar line at 121.6 nm. The solar flux penetrates to about 70 kilometers, where

8.2. ABSORPTION AND IONIZATION CROSS SECTIONS

it can photoionize the trace of nitric oxide present (see Table 8.1). The ionization source is primarily responsible for the creation of the D layer of the ionosphere. The cross section at the window is about $10^{-20}\,cm^2$, *Carver et al.*, 1977. See *Ogawa and Yamawaki*, 1970, for other window cross sections. As the cross section appears to be column density dependent, and the profile of the solar line is complicated, a complete understanding of the absorption of Lyman alpha will require additional measurements, *Nicolet and Peetermans*, 1980. Increased solar flux through other windows may also penetrate to lower altitudes and affect the D region, *Huffman et al.*, 1971; *Paulsen et al.*, 1972.

The **PHOTOIONIZATION CONTINUUM** of molecular oxygen begins at about 102.6 nm, although the presence of some vibrationally excited oxygen in the atmosphere leads to a small ionization yield at slightly longer wavelengths. The absorption rises to a broad maximum near 60 nm and then starts a steady drop toward the x-ray region, as shown in Figure 8.1. Near the threshold, there are a number of bands superimposed on the continuum. The photoionization cross section is also shown in the figure, as it is somewhat less than the photoabsorption cross section down to about 64 nm. Most of the bands are preionized, as shown.

The absorption spectrum of molecular oxygen is complicated on both sides of the ionization threshold, as reviewed by *Watanabe*, 1958. Cross section measurements made with continuum background light sources can map this structure, *Huffman et al.*, 1964a,b. Further improvements in the measurements available have been by *Matsunaga and Watanabe*, 1967, *Samson, et al.*, 1977b, and *Samson, et al.*, 1982. The review by *Conway*, 1988, is especially useful for atmospheric problems, as it is based on solar emission lines.

MOLECULAR NITROGEN

Although molecular nitrogen is the dominant species over much of the atmosphere, it does not absorb as strongly as oxygen in the wavelength region longward of its ionization threshold. In addition, nitrogen has the considerably larger dissociation energy of 9.759 eV, compared to 5.115 for oxygen, as shown in Table 8.1. Also, the solar flux at wavelengths shorter than the nitrogen photodissociation threshold of 127 nm is much less than the comparable region for oxygen beginning at 242 nm. There is thus no effective photodissociation process leading to the creation of atomic nitrogen that is comparable to molecular oxygen photodissociation. Atomic nitrogen does not play a major role as an ambient constituent in the upper atmosphere, but it is important in a number of reactions as reactant or product.

Strong molecular absorption bands begin at about 100 nm and continue to the first ionization threshold, as shown in Figure 8.2. Although only a few bands are shown, this is a very complicated group of Rydberg bands converging to the ionization threshold. The bands have narrow rotational lines, as seen in spectra, but at least some of these bands are predissociated. Since the solar spectrum consists of emission lines in this region, appreciable contribution to atmospheric absorption by molecular nitrogen requires the coincidence of a narrow solar line or multiplet with an absorption band usually composed of narrow rotational lines. The production of atomic nitrogen by photodissociation in the thermosphere at strong solar lines such as CIII at 97.7 nm is discussed by *Richards et al.*, 1981.

At wavelengths less than the ionization threshold at 79.6 nm, the strong bands continue until about 60 nm. The cross section continues to increase as additional ionization thresholds are reached, and the shape of the continuum is influenced by autoionization profiles of some of the bands. These Beutler-Fano profiles lead to shallow windows in the continuum at places. The photoionization yield becomes unity for wavelengths of about 65 nm and shorter. The peak cross section near 60 nm is about the same as the peak value for molecular oxygen. The cross section then decreases slowly from the peak toward the soft x-ray region.

Cross section measurements are available throughout the ultraviolet, but the resolution remains inadequate for the sharp bands. Cross sections for the molecular bands at wavelengths less than 100 nm have never been measured in sufficient detail to be used in a satisfactory way in absorption calculations. These cross sections are important for remote sensing of the ionosphere using 83.4 nm scattered radiation, as discussed by *Cleary et al.*, 1989, and also in Chapter 17.

Measured cross sections at relatively broad bandwidths and using continuum background sources are available, *Huffman et al.*, 1963; *Carter*, 1972. Improved measurements extending to the soft x-ray region have been reported by *Samson*, 1977a, 1987, and *Henke et al.*, 1982.

The possible products from photon absorption multiply greatly in the extreme ultraviolet and soft x-ray region. The products may be singly or multiply charged atomic or molecular ions. There may be inner shell excitation, and there are usually several important electronic states. The reader is

8.2. ABSORPTION AND IONIZATION CROSS SECTIONS

advised to look for data reviews that are concerned with atmospheric problems. For example, the data given in the massive and valuable review by *Gallagher et al.*, 1988, includes only the valence shell excitations and does not provide cross sections in sufficient detail for our use. This is especially apparent near the first ionization threshold. For atmospheric applications, the best initial source of information is a compilation designed for atmospheric use and keyed to solar emission features; such as *Conway* 1988.

ATOMIC OXYGEN

Atomic oxygen absorbs first at its resonance multiplet, which in spectroscopy is the allowed transition between the ground state and the excited state of lowest energy. The resonance line or multiplet is usually the strongest in emission and absorption spectra. For atoms such as oxygen having multiplet structure in many energy levels, there is a resonance multiplet of several lines rather than a single line.

For atomic oxygen, the resonance line absorption transition is

$$\mathrm{O}\, 2p^4\, {}^3P_{2,1,0} + h\nu(130\,nm) \longrightarrow \mathrm{O}\, 2p^3({}^4S_{3/2})3s\, {}^3S_1^o.$$

The resonance multiplet consists of several lines, with the strongest at 130.2, 130.5, and 130.6 nm, as a result of transitions between the fine structure levels. There are intense solar emission lines at the same wavelengths, and the resonance scattering of this major contributor to the UV radiance is a candidate for remote sensing use, as discussed in Chapter 15.

The 130 nm resonance lines are the first members of a Ryberg series converging to the first ionization threshold near 91 nm. There are other series each converging to one of the three ionization thresholds associated with the three lowest energy configurations of $\mathrm{O}^+(2p^3)$. See *Huffman et al.*, 1967a,b, for details about these series. The second member of the resonance series at a wavelength of about 102.5 nm overlaps the relatively strong solar Lyman beta line at roughly the same wavelength. Absorption lines of atomic oxygen and atomic nitrogen may influence measurements of the ionization continuum, *Huffman et al.*, 1966, and atmospheric composition measurements using the occultation of solar lines, *Huffman and Larrabee*, 1968.

The primary absorption of atomic oxygen in the atmosphere is in its photoionization continuum, which begins at about 91 nm, rises to a maximum at about 50 nm, and then slowly decreases toward the soft x-ray region (see

Figure 8.2). A few of the Rydberg series converging to excited states of O^+ are shown in the figure, but for additional details the references must be consulted.

Atomic oxygen is not stable at room temperatures and pressures. It can be used in the laboratory only by performing measurements rapidly, generally in a flowing afterglow or a beam. There are several methods used to produce it, but an associated problem for many measurements is to also determine the number density. The measurements must be made before it disappears by recombination to molecular oxygen, either in the gas phase or on the walls of the container, or by other reactions. The primary absorption of interest is in the extreme ultraviolet, where it is difficult to have suitable windows for an absorption cell and good light sources. The use of intense and available synchrotron light sources and a better understanding of the chemistry of oxygen in discharges and afterglows have resulted in good cross section measurements.

The atomic oxygen cross section curve is shown in Figure 8.2, which shows the continuum and a few of the discrete lines in the spectrum. For the most recent measurements, the reader is referred to *Samson and Pareek*, 1985, and *Angel and Samson*, 1988. As the energy increases toward shorter wavelengths in the extreme ultraviolet, the number of possible products of photoabsorption increase. The branching ratios, as known at this time, are given and discussed by *Conway*, 1988.

For atomic oxygen, O^{++} can be formed at wavelengths shorter than 24.8 nm, and for O^{+++} the corresponding threshold is 11.2 nm. However, the measurements of *Angel and Samson*, 1988, indicate that these processes are not dominant. At a wavelength of about 10 nm, they report partial photoionization cross sections of 1.01, 0.16, and about 0.001 Megabarns (or $10^{-18}\,cm^2$) for the single, double, and triple ionized atom products.

MINOR CONSTITUENTS

Which atmospheric constituents to refer to as minor is a matter of opinion, since many of the minor species in terms of number density cause major effects. For the region from the surface up to the thermosphere, the primary minor constituents include water vapor, carbon dioxide, argon, nitric oxide, and atomic nitrogen. There are many others in smaller concentrations, including the other common oxides of nitrogen, carbon monoxide, methane, hydrogen peroxide, sulfur compounds, and trace rare gases. The halogen substituted methanes and ethanes have assumed great importance recently as pollutants leading to ozone destruction.

Reviews of photon cross sections of minor species are in *Watanabe*, 1958; *Warneck*, 1988; *Brasseur and Solomon*, 1984; and *Rees*, 1989. The periodically updated reviews by *DeMore et al.*, 1990, provide authoritative information about pollutant molecules such as the fluorocarbons, as well as other species important in the stratosphere.

8.3 References

Anderson, G. P., F. X. Kneizys, E. P. Shettle, L. W. Abreu, J. H. Chetwynd, R. E. Huffman, and L. A. Hall, LOWTRAN 7 spectral simulations in the UV, *Conference Proceedings No. 254, Atmospheric Propagation in the UV, Visible, IR, and MM-Wave Region and Related Systems Aspects*, AGARD, NATO, 25-1 to 25-9, 1989 (Available from the National Technical Information Service, (NTIS), 5285 Port Royal Road, Springfield, VA 22161).

Angel, G. C. and J. A. R. Samson, Total photoionization cross sections of atomic oxygen from threshold to 44.3 Å *Phys. Rev. A, 38*, 5578–5585, 1988.

Banks, P. M. and G. Kockarts, *Aeronomy*, in two parts: A and B, Academic Press, 1973.

Berkowitz, J., *Photoabsorption, Photoionization, and Photoelectron Spectroscopy*, Academic Press, 1979.

Blake, A. J., An atmospheric absorption model for the Schumann-Runge bands of oxygen, *J. Geophys. Res., 84*, 3272, 1979.

Brasseur, G. and S. Solomon, *Aeronomy of the Middle Atmosphere*, Reidel, 1984.

Carter, V., High resolution N_2 absorption study from 750 to 980 Å *J. Chem. Phys., 56*, 4195, 1972.

Carver, J. H., H. P. Gies, T. J. Hobbs, B. R. Lewis, and D. G. McCoy, Temperature dependence of the molecular oxygen photoabsorption cross section near the H Lyman alpha line, *J. Geophys. Res., 82*, 1955, 1977.

Cleary, D. D., R. R. Meier, E. P. Gentieu, P. D. Feldman, and A. B. Christensen, An analysis of the effects of N_2 absorption on the O^+ 834 Å emission from rocket observations, *J. Geophys. Res., 94*, 17281–17286, 1989.

Conway, R. R., Photoabsorption and photoionization cross sections of O, O_2, and N_2 for photoelectron production calculations: A compilation of recent laboratory measurements, NRL Memorandum Report 6155, *Naval Research Laboratory, Washington, D.C.*, 1988.

DeMore, W. B., S. P. Sander, D. M. Golden, M. J. Molina, R. F. Hampson, M. J. Kurylo, C. J. Howard, and A. R. Ravishankara, *Chemical Kinetics and Photochemical Data for Use in Stratospheric Modeling, Evaluation Number 9*, JPL Publication 90-1, Jet Propulsion Laboratory, Pasadena, California, 1990.

Gallagher, J. W., C. E. Brion, J. A. R. Samson, and P. W. Langhoff, Absolute cross sections for molecular photoabsorption, partial photoionization, and ionic photofragmentation processes, *J. Phys. Chem. Ref. Data, 17*, 9–153, 1988.

Griggs, M., Atmospheric ozone, in *The Middle Ultraviolet: Its Science and Technology*, A. E. S. Green, editor, Wiley, 83–117, 1966.

Handbook of Geophysics and the Space Environment, A. S. Jursa, Scientific Editor, Air Force Geophysics Laboratory, NTIS Document Accession Number ADA 167000, 1985.

Henke, B. L., P. Lee, T. J. Tanaka, R. L. Shimabukuro, and B. K. Fujikawa, Low-energy x-ray interaction coefficients: photoabsorption, scattering, and reflection, *At. Data and Nucl. Data Tables, 27*, 1, 1982.

Herzberg, G., *Molecular Spectra and Molecular Structure I. The Spectra of Diatomic Molecules, 2nd Edition*, Van Nostrand, 1950.

Huber, K. P. and G. Herzberg, *Molecular Spectra and Molecular Structure IV. Constants of Diatomic Molecules*, Van Nostrand Reinhold, 1979.

Hudson, R. D., Critical review of ultraviolet photoabsorption cross sections for molecules of astrophysical and aeronomic interest, *Rev. Geophys. Space Phys., 9(2)*, May 1971; also Nat. Stand. Ref. Data Ser., NBS, NSRDS-NBS 38, 1971.

8.3. REFERENCES

Hudson, R. D., V. L. Carter, and J. A. Stein, An investigation of the effect of temperature on the Schumann-Runge absorption continuum of oxygen, 1580-910 Å, *J. Geophys. Res.*, *71*, 2295, 1966.

Huffman, R. E., Absorption cross-sections of atmospheric gases for use in aeronomy, *Canad. J. Chem.*, *47*, 1823, 1969.

Huffman, R. E., Chapter 22, Atmospheric emission and absorption of ultraviolet radiation, in *Handbook of Geophysics and the Space Environment*, A. Jursa, scientific editor, Air Force Geophysics Laboratory, NTIS Document Accession Number ADA 167000, 1985.

Huffman, R. E. and J. C. Larrabee, Effect of absorption by atomic oxygen and atomic nitrogen lines on upper atmosphere composition measurements, *J. Geophys. Res.*, *73*, 7419–7428, 1968.

Huffman, R. E., Y. Tanaka, and J. C. Larrabee, Absorption coefficients of nitrogen in the 1000-580 Å wavelength region, *J. Chem. Phys.*, *39*, 910, 1963.

Huffman, R. E., J. C. Larrabee, and Y. Tanaka, Absorption coefficients of oxygen in the 1060-580 Å wavelength region, *J. Chem. Phys.*, *40*, 356, 1964a.

Huffman, R. E., Y. Tanaka, and J. C. Larrabee, Nitrogen and oxygen absorption cross sections in the vacuum ultraviolet, *Disc. Faraday Soc.*, *37*, 159, 1964b.

Huffman, R. E., J. C. Larrabee, and Y. Tanaka, Influence of atomic oxygen absorption line series on cross section measurements, *Phys. Rev. Letters*, *16*, 1033, 1966.

Huffman, R. E., J. C. Larrabee and Y. Tanaka, New absorption spectra of atomic and molecular oxygen in the vacuum ultraviolet. I. Rydberg series from OI ground state and new excited O_2 bands, *J. Chem. Phys.*, *46*, 2213, 1967a.

Huffman, R. E., J. C. Larrabee, and Y. Tanaka, New absorption spectra of atomic and molecular oxygen in the vacuum ultraviolet. II. Rydberg series from OI $^1D_2$ and OI $^1S_0$ metastable states, *J. Chem. Phys.*, *47*, 4462–4471, 1967b.

Huffman, R. E., D. E. Paulsen, J. C. Larrabee, and R. B. Cairns, Decrease in D-region O_2 ($^1\Delta_g$) photoionization rates resulting from CO_2 absorption, *J. Geophys. Res.*, *76*, 1028, 1971.

Inn, E. C. Y. and Y. Tanaka, Absorption coefficients of ozone in the visible and ultraviolet regions, *J. Op. Soc. Am.*, *43*, 870–873, 1953.

Kirby-Docken, K., E. R. Constantinides, S. Babeu, M Oppenheimer, and G. A. Victor, Photoionization and photoabsorption cross sections of He, O, N_2, and O_2 for aeronomic calculations, *At. Data and Nucl. Data Tables*, *23*, 63, 1979.

Lean, J. L. and A. J. Blake, The effect of temperature on thermospheric molecular oxygen agsorption in the Schumann-Runge continuum, *J. Geophys. Res.*, *86*, 211, 1981.

Lewis, B. R., L. Berzins, and J. H. Carver, Oscillator strengths for the Schumann-Runge bands of $^{16}O_2$, *J. Quant. Spectrosc. Radiat. Transfer*, *36*, 209, 1986.

Matsunaga, F. M. and K. Watanabe, Total and photoionization coefficients and dissociation continua of rmO_2 in the 580–1070 Å region, em Sci. Light, 16, 191, 1967.

McDaniel, E. W., *Atomic Collisions, Electron and Photon Projectiles*, Wiley, 1989.

Metzger, P. H. and G. R. Cook, A reinvestigation of the absorption cross section of molecular oxygen in the 1050-1800 Åregion, *J. Quant. Spectros. Rad. Trans.*, *4*, 107, 1964.

Nicolet, M. and W. Peetermans, Atmospheric absorption in the O_2 Schumann-Runge band spectral range and photodissociation rates in the stratosphere and mesosphere, *Planet. Space Sci.*, *28*, 85, 1980.

Nicolet, M., S. Cieslik, and R. Kennes, Aeronomic problems of molecular oxygen photodissociation V. Predissociation in the Schumann-Runge bands of oxygen, *Planet. Space Sci.*, *37*, 427–458, 1989.

Ogawa, M. and K. R. Yamawaki, Absorption coefficients of O_2 at the Lyman-α line and of other O_2 transmission windows, *Appl. Optics*, *9*, 1709–1711, 1970.

8.3. REFERENCES

Paulsen, D. E., R. E. Huffman, and J. C. Larrabee, Improved photoionization rates of O_2 ($^1\Delta_g$) in the D-region, *Radio Science, 7*, 51, 1972.

Rees, M. H., *Physics and Chemistry of the Upper Atmosphere*, Cambridge U. Press, 1989.

Richards, P. G., D. G. Torr, and M. A. Torr, Photodissociation of rmN_2: A significant source for thermospheric atomic nitrogen, *J. Geophys. Res., 86*, 1495, 1981.

Samson, J. A. R. and P. N. Pareek, Absolute photoionization cross sections of atomic oxygen, *Phys. Rev. A, 31*, 1470, 1985.

Samson, J. A. R., G. N. Haddad, and J. L. Gardner, Total and partial photoionization cross sections of N_2 from threshold to 100 Å *J. Phys. B., 10*, 1749, 1977a.

Samson, J. A. R., J. L. Gardner, and G. N. Haddad, Total and partial photoionization cross-sections of rmO_2 from 100 to 800 Å, *J. Electron Spectros., 12*, 281, 1977b.

Samson, J. A. R., G. H. Rayborn, and P. N. Pareek, Dissociative photoionization cross sections of rmO_2 from threshold to 120 Å, *J. Chem. Phys., 76*, 393, 1982.

Samson, J. A. R., T. Masuoka, P. N. Pareek, and G. C. Angel, Total and dissociative photoionization cross sections of N_2 from threshold to 107 eV, *J. Chem. Phys., 86*, 6128, 1987.

Schiff, H. I., editor, Proceedings of the Symposium on Laboratory Measurements of Aeronomic Interest, *Canad. J. Chem., 47*, 1711–1941, No. 10, 1969 (also I.A.G.A. Symposium No. 8).

Schoen, R. I., Laboratory measurements of photoionization, photoexcitation, and photodetachment, *Canad. J. Chem., 47*, 1879-1900, 1969.

Stolarski, R. S. and N. P. Johnson, Photoionization and photoabsorption cross sections for ionospheric calculations, *J. Atm. Terr. Phys., 34*, 1691–1701, 1972.

Torr, M. R. and D. G. Torr, Ionization frequencies for solar cycle 21, Revised, *J. Geophys. Res., 90*, 6675, 1985.

Torr, M. R., D. G. Torr, R. A. Ong, and H. E. Hinteregger, Ionization frequencies for major thermospheric constituents as a function of solar cycle 21, *Geophys. Res. Lett., 6*, 771, 1979.

Trakhovsky, E., A. Ben-Shalom, U. P. Oppenheim, A. D. Devir, L. S. Balfour, and M. Engel, Contribution of oxygen to attenuation in the solar blind UV spectral region, *Appl. Opt., 28*, 1588–1591, 1989a.

Trakhovsky, E., A. Ben-Shalom, and A. D. Devir, Measurements of tropospheric attenuation in the solar blind UV spectral region and comparison with LOWTRAN 7 code, *SPIE, 1158*, 357–365, 1989b.

Warneck, P., *Chemistry of the Natural Atmosphere*, Academic Press, 1988.

Watanabe, K., Ultraviolet absorption processes in the upper atmosphere, *Advances in Geophysics, 5*, H. E. Landsberg and J. Van Mieghem, editors, Academic Press, 1958.

Yoshino, K., D. E. Freeman, J. R. Esmond, R. S. Friedman, and W. H. Parkinson, High resolution absorption cross-sections and band oscillator strengths of the Schumann-Runge absorption bands of isotopic oxygen, $^{16}O^{18}O$, at 79 K, *Planet. Space Sci., 37*, 419–426, 1989.

Chapter 9

Airglow

The naturally occurring emission sources in the atmosphere are potentially the basis for remote sensing methods. This chapter and the next two discuss radiance from airglow, aurora, and scattering radiance sources, respectively. Following these chapters, the ultraviolet background, or the total of all natural radiance sources in the atmosphere, is presented.

This chapter describes the UV airglow as now understood from ground, rocket, and satellite measurements and model calculations. Major satellite missions utilized are given in Section 5.5. More detail about specific emissions used for remote sensing applications, and modeling for this purpose, are given in Chapters 15, 16, and 17.

9.1 Excitation and measurement

The airglow as defined here is a consequence of one of the following excitation processes:

- Photoelectron collisional excitation
- Chemiluminescent reaction
- Ionic radiative recombination
- Photodissociative excitation
- Heavy particle collisional excitation (nonauroral)

These excitation processes are followed by radiative deexcitation, which is the airglow emission. Emission is usually in competition with collisional deexcitation, which results in no emission. The importance of radiative versus collisional deexcitation increases as the altitude increases and collisions become less frequent.

Most airglow emission processes vary with solar flux and thus are affected by solar flux variability. However, there have not been enough UV airglow measurements over several solar cycles to adequately model this variability. Thus, the intensities available at this time are as measured, and their radiance levels can be expected to vary with the solar cycle. Solar variability is mentioned in Section 7.2.

Airglow in this book is to be distinguished from scattering of incoming solar radiation, which is treated separately in Chapter 11. Auroral excitation, due to higher energy electron and proton collisions, is treated in Chapter 10.

The visible and near ultraviolet airglow can be observed from ground-based sites. The lower wavelength limit is at about 300 nm. For most of the emission of interest here, the observations must be made from sounding rockets or Earth-orbiting satellites. At the present time, there have been dozens of sounding rocket flights and a number of satellites, including OGO-4, S3-4, P78-1, and ASSI/San Marcos (see Section 5.5).

Airglow measurements determine the radiance, spectrum, altitude, and global extent of the emission. Most measurements are along the line of sight of a single pixel photometer or spectrometer. For interpretation with atmospheric models, the key measurement is the volume emission rate in *photons/cm^3* as a function of altitude, which can be obtained from sounding rockets flown through the emission. Measurements of airglow, however, give the apparent column emission rate in *photons $cm^{-2} sec^{-1} column^{-1}$*, or the apparent radiance in Rayleighs. These units are discussed in Chapter 3. Inversion techniques to convert to volume emission or to give the spatial structure in three dimensions are being developed: see, for example, *Solomon et al.*, 1984.

Airglow is discussed in virtually all books on aeronomy or the upper atmosphere. Features such as the oxygen green line (557.7 nm) and red line (630 nm) are routinely measured in airglow observatories around the world. There is a long history of these ground-based observations, and there is now a sizable volume of space observations. For the older literature, see *Chamberlain*, 1961. A report by *Barth*, 1965, and a shorter review, *Barth*, 1966, that appeared as detailed rocket and satellite measurements were beginning, continue to be useful.

9.2. DAY AIRGLOW

A recent review covering UV airglow has been given by *Meier*, 1991. This reference is especially recommended for further details concerning airglow spectroscopy, including recent modeling and measurements. A review of optical aeronomy in the years 1987–1990 includes recent airglow research in the U.S., *Solomon*, 1991.

9.2 Day airglow

The most important day airglow features are described here. Chapter 12 provides wavelengths in tabular form.

MOLECULAR NITROGEN BANDS: LBH, 2P, VK, BH

The **Lyman-Birge-Hopfield (LBH) bands** of molecular nitrogen occur in the day airglow from about 127 to 240 nm. With a moderate resolution of perhaps 0.5 nm, there are roughly 24 distinguishable peaks in the spectrum between 200 and 127 nm, with many of these peaks composed of several vibration bands. The emission is due to the transition

$$N_2\,(a\,^1\Pi_g) \rightarrow N_2\,(X\,^1\Sigma_g^+) + h\nu\,(LBH).$$

The bands are excited by photoelectrons resulting from solar extreme ultraviolet photoionizion in the upper atmosphere.

The intensities of the LBH bands are related to solar activity and to the solar zenith angle. The total band system usually has an intensity of several kiloRayleighs, with the strongest individual bands being several hundred Rayleighs. The peak of the emission as observed in the limb is at a tangent altitude of about 175 to 200 km. As observed from space, the intensities of the bands are modified by absorption due to atmospheric molecular oxygen in the Schumann-Runge continuum.

The upper state is metastable compared to fully allowed electric dipole transitions, and the reported lifetime varies from 56 to 115 μsec, with the lower one considered most likely, *Meier*, 1991. Collisional quenching of the state is the principal loss mechanism at lower altitudes.

As the LBH airglow is due to photoelectron collisions, the observed radiance and altitudinal distribution varies with the solar zenith angle. The radiance also varies with the solar cycle. The solar radiation responsible for the LBH bands is in the EUV, generally at wavelengths shorter than 50 nm. The incoming photon must ionize the atmospheric gas and leave the photoelectron produced with enough energy to excite nitrogen to the upper state of the LBH bands. Thus, long term models must include solar EUV variability.

The **Second Positive (2P) bands** of molecular nitrogen are present in the day airglow between roughly 300 and 400 nm, but they are not seen when viewing in the Earth direction from space, due to strong scattering from the troposphere and the albedo of the Earth surface. The bands are due to the transition

$$N_2\left(C\,^3\Pi_u\right) \rightarrow N_2\left(B\,^3\Pi_g\right) + h\nu\,(2P).$$

Note that the emission is between two excited states of the N_2 molecule. The (0,0) band at 337.1 nm may have an intensity of 25 kiloRayleighs in the horizontal direction at 150 km altitude. It has been used to remotely sense the photoelectron flux and the nitrogen vertical profile.

The **Vegard-Kaplan (VK) bands** of molecular nitrogen are similar to the second positive bands in that they can be observed only if the line of sight does not include solar scattering and albedo from near the surface. They are found at roughly 250 to 400 nm with intensities up to 10 kiloRayleighs or so for the individual bands. The transition is

$$N_2\left(A\,^3\Sigma_u^+\right) \rightarrow N_2\left(X\,^1\Sigma_g^+\right) + h\nu\,(VK).$$

The interpretation of this airglow and its use in remote sensing is complicated because the excitation of the upper state is partly through cascade from higher excited states of the molecule and because the $N_2\,A$ state is metastable. The radiative lifetime of the A state is about three seconds, which means that it can engage in a number of exothermic chemical reactions. These compete with radiative deexcitation in the atmosphere and complicate methods for the use of the bands for remote sensing, due to the need for accurate rate constants for quenching reactions of the A state.

The excess electronic excitation energy carried by the metastable $N_2\,A$ state has important atmospheric consequences, as it is involved in the excitation of the atomic oxygen green line at 557.7 nm, which is one of the principal airglow emissions. In addition, it is postulated to be a reactant in chemiluminescent reactions from missile plumes.

The **Birge-Hopfield (BH) bands** of molecular nitrogen have been found as weak emissions in the 125 to 98 nm wavelength region, with in-

9.2. DAY AIRGLOW

dividual bands no more intense than about 100 Rayleighs. The transition is as follows:
$$N_2\,(b\,^1\Pi_g) \rightarrow N_2\,(X\,^1\Sigma_g^+) + h\nu\,(BH).$$
These bands can be observed from space in the nadir direction, extending to wavelengths shorter than the LBH bands and in a region with strong atomic line emitters. The interpretation of these bands in the airglow is complicated because of multiple scattering.

ATOMIC OXYGEN LINES: 130.4, 135.6, 83.4 nm

The **130.4 nm** emission from neutral atomic oxygen is one of the most intense features of the UV background. The resonance multiplet, centered at 130.4 nm, is due to the transition $2p^4\,^3P_{2,1,0} - 2p^33s\,^3S_1^o$. The multiplet has intensities of 10 to 15 kiloRayleighs. It is excited by both solar scattering and photoelectron collisions, and its interpretation is complicated. As the primary radiance source under most conditions is scattering, it is discussed in Chapter 11.

The atomic oxygen airglow at **135.6 nm** is usually considered the best airglow emission for remote sensing of the ionosphere and of the atomic oxygen density in the thermosphere. It is primarily excited by photoelectron impact in the airglow and for most situations, the multiple scattering contribution is small. The transition is $2p^4\,^3P_{2,1,0} - 2p^33s\,^5S_2^o$. It is found at intensities of about one kiloRayleigh, which makes it more intense than the N_2 LBH bands found in the nearby wavelength region.

The transition consists of two lines, the dominant transition at 135.6 to the $J = 2$ level of the ground state, and a weaker transition to the $J = 1$ level occurring at 135.9 nm. In practice, and for most available measurements, both of these lines are meant when the 135.6 nm line is mentioned. Another complication in the dayglow and aurora is the N_2 LBH (3,0) band located at 135.4 nm. It is not possible with typical wavelength resolutions available today to isolate the oxygen emission from the LBH band, and since the band is about one-fifth the intensity of the 135.6 nm transition, a correction is necessary.

There are many other emission lines from neutral atomic oxygen in the day airglow. Lines of sufficient intensity to be readily observable are at 164.1, 117.3, 115.2, 104.0, 102.7, and 98.9 nm.

Emission lines from singly ionized atomic oxygen appear in the day airglow in the EUV beginning at **83.4 nm**. The energy levels involved are

$2s^22p^3\,^4S^o_{3/2} - 2s2p^4\,^4P_{5/2,3/2,1/2}$. This is also the strongest line from O^+ in the airglow, with an intensity of up to about 600 Rayleighs. A number of weaker lines from singly ionized atomic oxygen have been identified or possibly identified in the day airglow down to about 43 nm (see Chapter 12).

Note that the lines and energy levels for singly ionized atomic oxygen, O^+, may be referred to in tables as OII, a terminology common in atomic spectroscopy. Using this nomenclature, the neutral atom levels are referred to as OI; doubly charged atomic oxygen would be OIII, etc. Similar terminology may be used for N and other atoms.

ATOMIC NITROGEN LINES: 149.3, 120, 214.3 nm

Proceeding into the ultraviolet toward shorter wavelengths, emission from neutral atomic nitrogen begins in the day airglow at **149.3 nm**, which is due to the transition $2p^3\,^2D^o_{5/2,3/2} - 2p^23s\,^2P_{3/2,1/2}$. The emission intensity may be several hundred Rayleighs.

The brightest line from neutral atomic nitrogen is at **120 nm**, which can be up to 1.5 kiloRayleigh. This line is close to the extremely intense (up to 20 kiloRayleigh) hydrogen Lyman alpha line at 121.6 nm, which makes accurate measurement of it difficult. The transition involved is $2p^3\,^4S^o_{3/2} - 2p^23s\,^4P_{5/2,3/2,1/2}$. Another strong line at 113.4 nm can be up to 400 Rayleighs, and there are a number of additional lines identified or tentatively identified in the FUV and EUV dayglow down to about 85 nm. These are shown in the tables in Chapter 12.

The strongest dayglow features involving transitions of singly ionized atomic nitrogen are at **214.3**, 108.5, and 91.6 nm. The remaining lines, primarily in the EUV, are much weaker. The emission at 214.3 was confused for many years with the (1,0) NO gamma band. The atomic transition is now known to occur both in the dayglow and in the aurora. The transition is $2s^22p^2\,^3P_{0,1,2} - 2s2p^3\,^5S_2$, and intensities of around 300 Rayleigh have been found. The 108.5 and 91.6 nm emission features also involve inner subshell transitions and have intensities in the dayglow of about 500 and 200 Rayleighs, respectively.

A rocket spectrum of the day airglow in the FUV is shown in Figure 9.1. This spectrum is due to *Eastes et al.*, 1985, and it is at a resolution of 0.35 nm. Most reported measurements are at larger resolutions. A bandwidth of at least 1 nm is needed to find out much about the airglow spectrum in any

9.3. NIGHT AIRGLOW

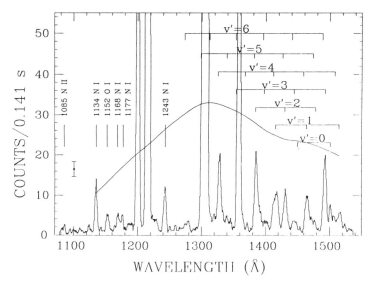

Figure 9.1: Day airglow spectrum in the FUV, showing nitrogen LBH emission (*Eastes et al.*, 1985, used with permission)

detail. A satellite spectrum in the FUV is shown in Figure 9.2, *Huffman et al.*, 1980).

Several EUV emission features have been mentioned in the above discussion. A complete listing and more details of the P78-1 measurements is given by *Chakrabarti et al.*, 1983. A listing of the most important airglow wavelengths and assignments is given in Chapter 12 on UV backgrounds.

9.3 Night airglow

MOLECULAR OXYGEN HERZBERG BANDS

The **molecular oxygen Herzberg (Hz) bands**, and similar, weaker oxygen emission bands, are found between about 390 and 250 nm in the night airglow. The Herzberg bands are located in a fairly narrow altitude region near 95 km. The transition for the most prominent band system, the Herzberg I system, is as follows: $A\,^3\Sigma_u^+ - X\,^3\Sigma_g^-$. In addition, the Herzberg II

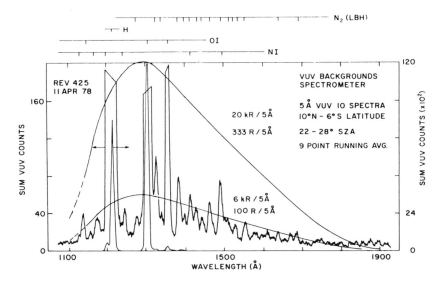

Figure 9.2: Day airglow from S3-4 satellite in FUV (*Huffman et al.*, 1980)

and Chamberlain bands have been found in this region. The Herzberg band upper state is metastable, with the lifetimes varying with the vibrational state. Mean lifetimes between about 0.1 and 0.5 sec have been measured.

The nadir viewing intensity due to molecular oxygen emission for the total 390 to 250 nm wavelength region is from 300 to 1000 Rayleighs. Individual bands may be up to about 10 Rayleighs. A spectrum of the night airglow in the MUV is shown in Figure 9.3. This measurement is from the S3-4 satellite *Eastes et al.*, 1992.

The reaction responsible for the emission is the radiative recombination of ground state oxygen atoms:

$$O\,(^3P) + O\,(^3P) \rightarrow O_2\,(A\,^3\Sigma_u^+).$$

The A state then undergoes a transition to the ground state with emission of a photon:

$$O_2\,(A\,^3\Sigma_u^+) \rightarrow O_2\,(X\,^3\Sigma_g^-) + h\nu \text{ (Herzberg)}.$$

The emission is therefore dependent on the atomic oxygen number density and can be used to obtain the atomic oxygen column or volume density.

9.3. NIGHT AIRGLOW

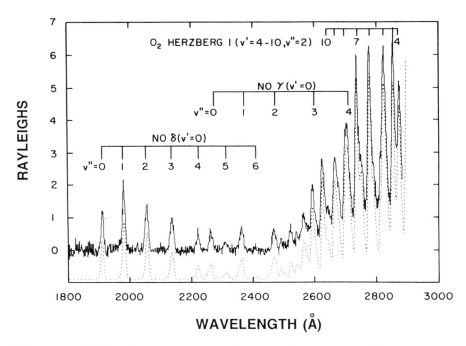

Figure 9.3: Night airglow spectrum from the S3-4 satellite (*Eastes et al.*, 1992)

The excited A state molecular oxygen can be quenched by collision. In addition, a competing, nonemitting reaction path involving the same reactants is three body recombination:

$$O + O + M \to O_2 + M.$$

In this reaction, all of the reactants and products are in their ground states, and the third body M can in principle be any atmospheric atom or molecule. Since the reaction rate at any altitude is proportional to the oxygen atom concentration squared times the density of M, the rate decreases more rapidly with increase in altitude than two body radiative recombination, and at some altitude the two body reaction will become more important.

In three body recombination, the third body, which usually undergoes no change in its internal energy, allows the process to proceed more rapidly by acquiring some of the excess kinetic and potential energy released in the reaction. This reaction rate can be described as a three body reaction, and further consideration of the dynamics of the process is outside our scope here.

Oxygen molecular emissions in the night airglow can be measured from ground sites also, and this work can be traced through a recent paper on measurements and models by *Stegman and Murtagh*, 1991.

NITRIC OXIDE GAMMA AND DELTA BANDS

The **nitric oxide gamma (γ) and delta (δ) bands** are produced by chemiluminescent reactions in the night airglow. The upper states involved are $A\,^2\Sigma^+$ for the gamma bands and $C\,^2\Pi$ for the delta bands. The lower level is the ground state of NO, $X\,^2\Pi$. The bands are also seen in fluorescence in daytime limb observations and at twilight (see Chapter 11).

The region of the atmosphere leading to this emission extends over a larger altitude range than for the Herzberg bands of oxygen. However, the same considerations of quenching and three body recombination apply.

The NO γ and δ bands are found in the 270 to 190 nm wavelength region. The emission is weak in the nightglow, with the total delta band emission between 20 and 30 Rayleighs and no individual band greater than about 8 Rayleigh. The gamma bands are even weaker, with a total emission of under 10 Rayleigh and no band greater than 4 Rayleighs in column emission rate.

Despite the weak intensity, they have been used for obtaining nitrogen atom and nitric oxide concentrations at night. The night airglow spectrum of these bands from the S3-4 satellite is shown in Figure 9.3.

9.3. NIGHT AIRGLOW

Extreme ultraviolet night airglow emissions, primarily from atomic lines, have been reported by *Chakrabarti et al.*, 1984, as measured by the P78-1 satellite. Observed emission lines in nm and approximate intensities in the EUV are OI 102.7, HI 102.5 (2.3 R); OI 98.9 (2.6 R); OI 91.1 (35 R); OII 83.4 (0.7 R); and HeI 58.4 (0.4 R).

Night midlatitude and equatorial spectra in the FUV and the MUV are shown in Figures 9.4 and 9.5, which were obtained from the S3-4 satellite. The emissions discussed in this section can be seen, as well as a scattered contribution from the F-region equatorial airglow emitting at higher altitudes (to be discussed next). The UV spacecraft glow, covered in Section 5.4, dominates in the north midlatitude panels.

EQUATORIAL IONOSPHERIC EMISSIONS

One of the principal discoveries of OGO-4, launched in 1968, was ultraviolet radiation occurring on either side of the Earth's magnetic dip equator in the early evening hours, *Barth and Schaffner*, 1970, and *Hicks and Chubb*, 1970. This radiation increase was first found for the atomic oxygen emission multiplets at 130.4 and 135.6 nm. It was named the tropical UV airglow, which is still a suitable name. One particularly striking early observation is in the images of the Earth at FUV wavelengths obtained from the moon on the Apollo program, as clearly seen in Figure 2.3. It has been found also at EUV wavelengths, *Paresce*, 1979; *Chakrabarti*, 1984.

Subsequent investigation has tied the emission to the radiative recombination of atomic oxygen ions and electrons in the F-region of the ionosphere; see, for example, *Anderson*, 1972. It is also associated with the "anomalous" ionosphere, or the Appleton anomaly, seen at these latitudes day and night. The emission has the potential of being valuable for remote sensing of the ionosphere.

This emission is found in two belts located 10 to 15 degrees on either side of the magnetic dip equator, which is the line around the Earth near the geographic equator where the Earth's magnetic field is parallel with the surface. The maximum emission rate is usually different for the northerly and southerly belts, with either being the most intense. Maximum intensities can range up to several hundred Rayleighs. The locations and strengths vary over time.

After early consideration of several emission sources, the consensus now is that this airglow emission is due to the radiative recombination of singly

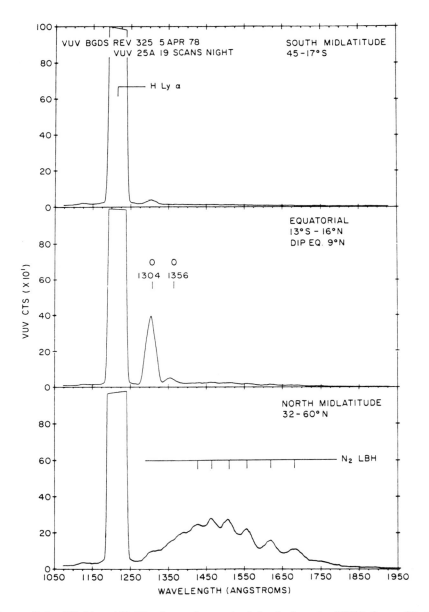

Figure 9.4: Night midlatitude and equatorial airglow in FUV from S3-4 satellite(*Huffman et al.*, 1980)

9.3. NIGHT AIRGLOW

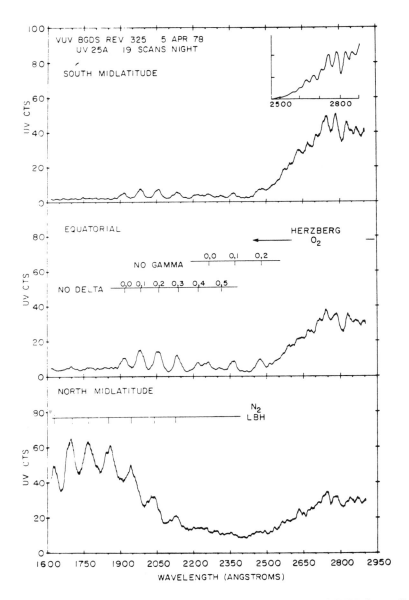

Figure 9.5: Night midlatitude and equatorial airglow in MUV from S3-4 satellite(*Huffman et al.*, 1980)

ionized atomic oxygen ions and electrons, as shown:

$$O^+ + e \to O^* \to O + h\nu.$$

The excited product atomic oxygen, O*, can be in any one of the huge number of electronic states of neutral atomic oxygen, since the energy made available by the radiative recombination is at least the ionization energy. For the lowest state of the ion, this is 13.6 ev. This is a large amount of energy for an airglow and leads to the formation of many excited states. Cascading, with weak emission at longer wavelengths, results in additional population of the upper states of the resonance multiplets of the triplet and quintet excited states, which emit at 130.4 and 135.6 nm, respectively. Final radiative decay is to the atomic oxygen ground state.

In addition to the 130.4 and 135.6 nm emission lines, other atomic oxygen emission lines observed in the tropical nightglow are at 297.2, 102.6, and 98.9 nm. The radiative recombination continuum associated with the kinetic energies of the recombining electrons is seen extending a short distance below 91.1 nm, which is the ionization threshold for ground state atomic oxygen. It has a total intensity of about 30 Rayleighs.

There are other emission sources that could be classified as airglows occasionally found, such as **heavy particle collisional excitation**, as reviewed by *Tinsley*, 1981. These sources, while interesting in themselves, may interfere with the use of the normal airglow for remote sensing, since they may emit at the same wavelength. However, after they are better characterized, they may themselves be incorporated into remote sensing methods as a proxy for ionospheric currents and fields with which they are associated.

9.4 References

Anderson, D. N., Theoretical calculations of the F-region tropical ultraviolet airglow intensity, *J. Geophys. Res.*, 77, 4782–4789, 1972.

Barth, C. A., *Ultraviolet Spectroscopy of Planets*, NASA Jet Propulsion Laboratory, Tech. Rpt. 32-822, Pasadena, California, 1965.

Barth, C. A., The ultraviolet spectroscopy of planets, Chapter 10 in *The Middle Ultraviolet: Its Science and Technology*, A. E. S. Green, editor, Wiley, 1966.

9.4. REFERENCES

Barth, C. A. and S. Schaffner, OGO-4 spectrometer measurements of the tropical ultraviolet airglow, *J. Geophys. Res.*, *75*, 4299–4306, 1970.

Chakrabarti, S., EUV (800–1400 Å) observations of the tropical airglow, *Geophys. Res. Lett.*, *6*, 565, 1984.

Chakrabarti, S., F. Paresce, S. Bowyer, R. Kimble, and S. Kumar, The extreme ultraviolet day airglow, *J. Geophys. Res.*, *88*, 4898–4904, 1983.

Chakrabarti, S., R. Kimble, and S. Bowyer, Spectroscopy of the EUV (350–1400 Å) nightglow, *J. Geophys. Res.*, *89*, 5660–5664, 1984.

Chamberlain, J. W. *Physics of the Airglow and Aurora*, Academic Press, 1961.

Eastes, R. W., P. D. Feldman, E. P. Gentieu, and A. B. Christensen, The ultraviolet dayglow at solar maximum 1. Far UV spectroscopy at 3.5 Å resolution, *J. Geophys. Res.*, *90*, 6594–6600, 1985.

Eastes, R. W., R. E. Huffman, and F. J. LeBlanc, NO and O_2 ultraviolet nightglow and spacecraft glow from the S3-4 satellite, *Planet. Spa. Sci.*, *40*, 481–493, 1992.

Hicks, G. T. and T. A. Chubb, Equatorial aurora/airglow in the far ultraviolet, *J. Geophys. Res.*, *75*, 6233–6248, 1970.

Huffman, R. E., F. J. LeBlanc, J. C. Larrabee, and D. E. Paulsen, Satellite vacuum ultraviolet airglow and auroral observations, *J. Geophys. Res.*, *85*, 2201–2215, 1980.

Meier, R. R., Ultraviolet spectroscopy and remote sensing of the upper atmosphere, *Space Sci. Rev.*, *58*, 1–185, 1991.

Paresce, F., EUV observation of the equatorial airglow, *J. Geophys. Res.*, *84*, 4409–4412, 1979.

Solomon, S. C., Optical aeronomy, in section entitled *Solar-Planetary Relationships: Aeronomy*, M. H. Rees, editor, *Rev. Geophys., Supplement, April 1991*, 1089–1109, 1991.

Solomon, S. C., P. B. Hays, and V. J. Abreu, Tomographic inversion of satellite photometry, *Appl. Optics*, *23*, 3409–3414, 1984.

Stegman, J. and D. P. Murtagh, The molecular oxygen band systems in the U.V. nightglow: Measured and modelled, *Planet. Spa. Sci.*, *39*, 595–609, 1991.

Tinsley, B. A., Neutral atom precipitation—A review, *J. Atmos. Terres. Phys.*, *43*, 617, 1981.

Chapter 10

Aurora

Light emission from the aurora in the polar regions of the Earth has excited interest for centuries. It is no less exciting when imaged from space. The arctic and antarctic poles are crowned with shimmering halos, or ovals, that follow the sun as the Earth rotates beneath.

The emission from the aurora is due to collisions of energetic electrons and protons with the atmospheric constituents. The more abundant electrons have an energy distribution peaking between 1 and 10 keV. The energetic particles are the result of complicated interactions of the solar wind with the Earth's magnetic field.

In the last decade, it has become possible to routinely image the aurora from space both in visible and ultraviolet wavelengths, as described in Chapter 16. Auroral imaging in the ultraviolet is a key element in global space weather systems now under development, as covered in Chapter 17.

This chapter is an introduction to aurora and their UV spectra. For more details on all aspects of the aurora, see the books by *Vallance Jones*, 1974, and *Chamberlain*, 1961. Conference proceedings providing the reader with more detailed information are available: *McCormac*, 1967; *Deehr and Holtet*, 1981; *Meng et al.*, 1991. Excellent scientific introductions to the aurora and related areas of solar-terrestial physics are given in articles in the *Handbook of Geophysics and the Space Environment* by *Whalen*, 1985; *O'Neil and Picard*, 1985; *Burke et al.*, 1985, and *Rich*, 1985.

A true appreciation of the aurora is difficult to acquire without personal observation of a display in the polar night sky. It has even been said that a person should not attempt a theory or model, or even a measurement, of the aurora without witnessing at least one. Several books and articles of a popular nature can help the reader to acquire a feel for what the aurora looks and acts like. Recommended are articles by *Akasofu*, 1965, 1981, 1989, and books by *Eather*, 1980, and *Brekke and Egeland*, 1983.

10.1 The auroral oval

The principal stuctural feature in UV global images of polar regions from space is the auroral oval. It is usually observed as an uneven band of emission between about 70 to 75 degrees in latitude and centered roughly on the magnetic pole. Although the concept of the oval had been previously suggested, all-sky camera images obtained during the International Geophysical Year in 1957–58 at a number of global locations demonstrated that the oval was always present at some latitude. This work is described by *Feldstein and Starkov*, 1970, and *Feldstein and Galperin*, 1985.

The auroral oval is correlated with the magnetic index K_p, being smaller and closer to the pole for low values and extending toward the equator for larger values. The auroral oval for $K_p = 3$, a generally used average value, is shown in Figure 10.1. This figure is in magnetic coordinates, which are standard for auroral studies. Auroral activity is greater around the time of the maximum of the 11 year solar cycle, when geomagnetic activity also is higher.

Aurora are a part of what is now understood to be a complex structure of electric and magnetic fields, together with associated currents of charged particles, surrounding the Earth. The concept is of the Earth as a body with its own magnetic field sitting like a rock in a stream within the flow of the solar wind and the associated solar fields. The solar wind is a varying stream of energetic particles coming from the sun. This flow of particles has associated with it electric and magnetic fields. The auroral regions represent the places on the globe where high energy electrons and protons excited by the solar wind and trapped on magnetic field lines reach altitudes where they collide with atmospheric atoms and molecules.

There are many different types of auroral forms. The auroral oval concept was developed for discrete auroral forms such as arcs, which are the brightest features. There are also diffuse, or continuous, aurora, which are less well modeled by the auroral oval. These auroral forms are usually not as bright as discrete aurora, but they may well represent a larger energy input to the atmosphere, as they extend over a larger spatial extent.

The auroral oval region is continually being distorted by magnetic sub-

10.1. THE AURORAL OVAL

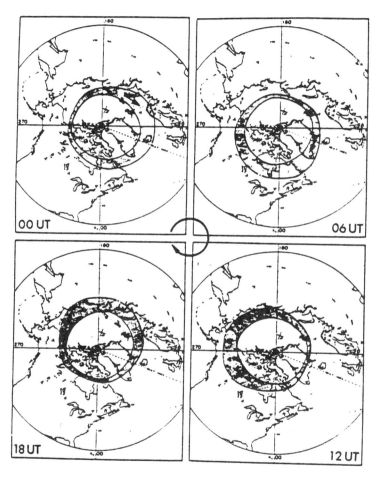

Figure 10.1: Northern auroral oval for four universal times (*Whalen*, 1985)

storms that alter the appearance and emission intensity of the aurora. This behavior is another contributor to the complexity of the auroral region. The auroral regions also have the conjugate property. When mapped in magnetic coordinates, it is found that the displays in the northern and southern auroral zones are closely similar.

The auroral oval location and intensity reflects the state of solar activity and sets limits on the locations and electron densities in the ionosphere. The electromagnetic structure of the Earth's upper atmosphere has been compared to a giant cathode ray tube generating streams of high energy charged particles that impinge on the atmospheric gases, which take the place of the phosphor on a CRT screen and emit to form the display. This display can be used for remote sensing of the ionosphere and magnetosphere, similar to the way that images of clouds from space are used for predicting surface weather.

A recent simplified description of theories of the aurora, including its particles, fields, currents, and emissions, has been given by *Akasofu*, 1989. More detailed treatments of the solar-terrestrial physics involved in the aurora can be found in books on the ionosphere and the magnetosphere. In addition to the sources previously given in this chapter, there are books by *Kelley*, 1989; *Akasofu and Chapman*, 1972; and *Akasofu and Kamide*, 1987. The text by *Davies*, 1989, includes a good introduction to solar-terrestrial relationships.

The energetic particles lose their energy at ionospheric altitudes, with higher energy particles penetrating deeper into the atmosphere. About 34 eV is lost by an electron, on the average, at each collision with an atmospheric atom or molecule. As a rough rule for electrons, 0.1 to 1 keV particles penetrate to F region altitudes of about 160 to 300 km; 1 to 40 keV particles are to E region altitudes of about 90 to 160 km; and electrons with energies greater than 40 keV penetrate to D region altitudes of 70 to 90 km. Thus, most of the auroral excitation takes place in the E region, since typical electron energies are 1 to 10 keV. Auroral light emission reaches a maximum in the E region from about 120 to 150 km.

10.2 Auroral spectroscopy

Light emission from the aurora in the ultraviolet must be measured from rockets or satellites, with the exception of the 300 to 400 nm region, which can be observed from the ground. Rocket spectra are especially suitable

10.2. AURORAL SPECTROSCOPY

for giving the altitudinal distribution and limb viewing intensities. Satellites give a global view, but there have been fewer experiments. A recent overview of auroral spectroscopy, including the ultraviolet region, has been given by *Vallance Jones*, 1991.

The discussion here provides auroral UV intensities as reported. It must be understood that the auroral UV radiances are highly variable, and the values given are perhaps most useful for relative intensities. There is much that is unknown about the spectra of aurora. The values are for nadir viewing, and there are virtually no limb viewing measurements. Differences in spectra for diffuse and arc, day and night, local time, solar activity, etc., remain to be investigated more thoroughly. The emission bands are generally the same as in the airglow, and reference should be made to their discussion in Chapter 9 for energy levels.

Auroral emission in the far ultraviolet from satellite photometers was reported by *Joki and Evans*, 1969, and by *Clark and Metzger*, 1969. A large data set of far ultraviolet auroral observations from the OGO-4 satellite was published by *Chubb and Hicks*, 1970. The ion cells used for detectors on OGO-4 allowed hydrogen Lyman alpha and nitrogen Lyman-Birge-Hopfield bands to be observed separately. Photometer measurements with bandwidths of between 12 and 17 nm and centered at 125, 134, 150, and 170 nanometers were obtained from S3-4 and compared to spectral observations on the same satellite *Huffman et al.*, 1980. Auroral ultraviolet spectra observed in the nadir direction from satellite altitudes has been reported from OGO-4 by *Gerard and Barth*, 1976; from P78-1 by *Paresce et al.*, 1983; and from S3-4 by *Ishimoto et al.*, 1988.

Auroral spectral features and intensities are given in a series of three tables covering the NUV, MUV and FUV, and EUV, respectively. Table 10.1 is based on *Vallance Jones*, 1974, who gives extensive examples of auroral spectra from the infrared through the far ultraviolet. The ultraviolet spectra given there are based on rocket data, and intensities are given based on a 557.7 nm (green line) intensity of 100 kiloRayleighs. This is considered a reasonably bright aurora, or International Brightness Class (IBC) III. The IBC is a logarithmic scale based on the brightness of the 557.7 nm atomic oxygen emission in kiloRayleighs.

Table 10.2 gives MUV and FUV emission features and intensities based on S3-4 satellite data both as reported by *Ishimoto et al.*, 1988, and from additional data from the author's laboratory at the Geophysics Directorate, Phillips Laboratory. It is difficult to base MUV and FUV auroral intensities on an IBC class, because of the few measurements to date, so the table

Table 10.1: NUV auroral intensities. See text for details.

Transition	λ (nm)	B (kR)	Comment
N_2^+ 1N (0,0)	391.4	60	1N: $B^2\Sigma_u^+ - X^2\Sigma_g^+$
N_2^+ 1N (1,1)	388.4	5	
N_2 2P (0,2)	380.5	20	2P: $C^3\Pi_u - B^3\Pi_g$
N_2 2P (1,3)	375.5	15	
N_2 2P (0,1)	357.7	30	
N_2 2P (1,2)	353.7	15	
N_2 2P (0,0)	337.1	20	
N_2 2P (1,0)	315.9	3	
N_2 2P (2,0)	297.0		blend w/O I
O I	297.2	2	$2p^4\ ^3P - 2p^4\ ^1S$

represents averages obtained in the April to September, 1978, time period. The measurements are based on nadir viewing from approximately 200 to 250 km altitudes.

Table 10.3 shows auroral EUV spectra and nadir intensities as reported by *Paresce et al.*, 1983. The measurements are from the P78-1 satellite at an altitude of 600 km. This satellite also found measurable emission in the zenith, or upward, direction for multiple scattered emissions such as 130.4, 121.6, 120.0, 104.0, 98.9, and 83.4 nm.

The auroral emission lines and bands given in Tables 10.1, 10.2, and 10.3 have been measured at night. The daytime aurora is somewhat different from the nighttime aurora, but the same emissions are generally dominant. For nadir and Earth-direction viewing of the auroral zone in the daytime, radiance from airglow and scattering sources may present a problem. In the NUV and MUV, scattering radiance should prevent auroral emission measurements. In the FUV and EUV, however, the day auroral emission is stronger or comparable to the airglow radiance sources important in these regions. Thus, the auroral oval can be studied in the daytime from space. The relative intensities of the airglow and aurora at auroral latitudes can be seen in Figures 12.4 through 12.7. The global auroral imaging discussed in Chapter 16 also depends on the relatively small background emission from airglow and scattering in the FUV.

The nadir viewing spectrum from the S3-4 satellite of the night aurora in the FUV and MUV is shown in Figure 10.2, *Ishimoto et al.*, 1988. This spectrum was built up from a number of passes of the satellite over diffuse

10.2. AURORAL SPECTROSCOPY

Table 10.2: MUV and FUV auroral intensities. See text for details.

Transition	λ (nm)	B (kR)	Comment
N_2 VK (0,6)	276.1	0.4	VK: $A^3\Sigma_u^+ - X^1\Sigma_g^+$
N_2 VK (0,5)	260.4	0.2	
N_2 VK (0,4)	246.2	0.2	
N_2 VK (1,4)	237.8	0.1	
N_2 VK (0,3)	233.3	0.1	
N_2 VK (1,3)	225.8	0.9	
N II	214.3	0.2	$2s^22p^2\ ^3P - 2s2p^3\ ^5S^o$
N_2 LBH $\Delta v = 8$	205.-	0.1	LBH: $a^1\Pi_g - X^1\Sigma_g^+$
N_2 LBH $\Delta v = 7$	195.-	0.2	
N_2 LBH $\Delta v = 6$	184.-	0.2	
N_2 LBH $\Delta v = 5$	176.-	0.4	
N_2 LBH $\Delta v = 4$	167.-	0.3	
N_2 LBH $\Delta v = 3$	160.-	0.2	
N I	149.3	0.4	$2p^3\ ^2D^o - 3s\ ^2P$
N_2 LBH (1,1)	146.4	0.05	
N_2 LBH (2,1)	143.0	0.03	
N_2 LBH (1,0)	141.6	0.07	also LBH (4,2)
N_2 LBH (2,0)	138.4	0.08	also LBH (5,2)
O I	135.6	0.6	$2p^4\ ^3P^o - 3s\ ^5S$
N_2 LBH (3,0)	135.4	0.08	est., blend O I
N_2 LBH (4,0)	132.5	0.14	
O I	130.4	15.	$2p^4\ ^3P^o - 3s\ ^3S$
N_2 LBH (5,0)	129.8	.	blend O I
N_2 LBH (6,0)	127.3	0.08	
H Lyman α	121.6	3.	$1s\ ^2S^o - 2p\ ^2P$
N I	120.0	0.5	$2p^3\ ^4S^o - 3s\ ^4P$
N I	116.8	0.1	$2p^2\ ^2D^o - 3d\ ^2F$
O I	115.2	0.01	$2p^4\ ^1D - 3s'\ ^1D^o$
N I	113.4	0.09	$2p^3\ ^4S^o - 2p^4\ ^4P$

Table 10.3: EUV auroral intensities. See text for details.

Transition	λ (nm)	B (kR)	Comment
N$_2$ BH (1,5)	111.0	0.2	BH: $b^1\Pi_u - X^1\Sigma_g^+$
N II	108.5	2.1	$2p^2\ ^3P - 2p^3\ ^3D^o$
O I	104.0	0.1	$2p^4\ ^3P - 4s\ ^3S^o$
O I	102.7	1.1	$2p^4\ ^3P - 3d\ ^3D^o$
N$_2$ BH (1,1)	100.9	0.5	
O I	98.9	2.0	$2p^4\ ^3P - 3s'\ ^3D^o$
N$_2$ BH (3,1)	96.5	0.2	
N I	95.3	0.4	$2p^3\ ^4S^o - 3d\ ^4P$
N II	91.6	0.2	$2p^2\ ^3P - 2p^3\ ^3P^o$
O II	83.4	0.8	$2p^3\ ^4S^o - 2p^4\ ^4P$
O II	79.7	0.04	$2p^3\ ^2P^o - 2p^4\ ^2D$
O II	71.8	0.07	$2p^3\ ^2D^o - 2p^4\ ^2D$
O II	61.7	0.06	$2p^3\ ^2D^o - 3s\ ^2P$
He I	58.4	0.08	$1s^2\ ^1S - 2p\ ^1P$
O II	53.8	0.2	$2p^3\ ^4S^o - 3s\ ^4P$
O II	51.5	0.04	$2p^3\ ^2P^o - 3d\ ^2D, P$
O II	48.2	0.04	$2p^3\ ^2D^o - 3d\ ^2F, P, D$
O II	43.0	0.06	$2p^3\ ^4S^o - 3d\ ^4D$
He II (est.)	30.4	0.01	$1s\ ^2S - 2p\ ^2P^o$

10.2. AURORAL SPECTROSCOPY

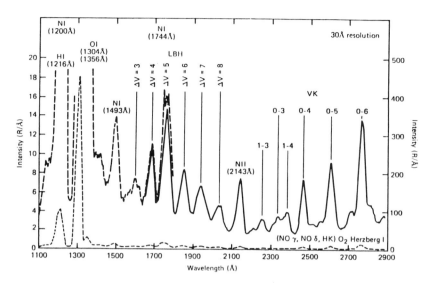

Figure 10.2: Auroral spectra in the FUV and MUV from the S3-4 satellite (*Ishimoto et al.*, 1988)

aurora, as the spectrometer required about 22 seconds to complete a spectral scan. Spectrometers with high sensitivity and good spatial resolution will be required to measure simultaneously the spectral, spatial, and temporal variations in the aurora.

The imaging of the aurora from space satellites is discussed in detail in Chapter 16. As an example of this method of obtaining spectral information about the aurora, Figure 10.3 illustrates multispectral images obtained by the Polar BEAR satellite, *Huffman et al.*, 1989. Note that the images are slightly different for the three emission features: OI, 130.4 nm; N_2, LBH band at 154.4 nm; and N_2^+, 1st negative band, 391.4 nm. While the general shape and the location of the oval, taken as a whole, are the same for the three wavelength bands, the detailed structure is not. These images reveal complexity in the auroral formation process that will have to be addressed in future auroral research. In particular, some concern about conclusions drawn from wide bandwidth filters would appear to be warranted.

As auroral spectroscopy improves, additional minor excitation mechanisms may be revealed by changes in vibration band intensity distributions, *Eastes and Sharp*, 1987, or by discovery of emission sources effectively hidden near other features in lower resolution spectra. An emission feature found

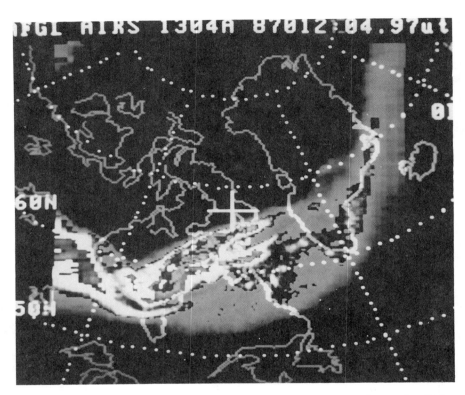

Figure 10.3: Simultaneous multispectral auroral images from the Polar BEAR satellite, OI, 130.4 nm (*Huffman et al.*, 1989)

10.2. AURORAL SPECTROSCOPY

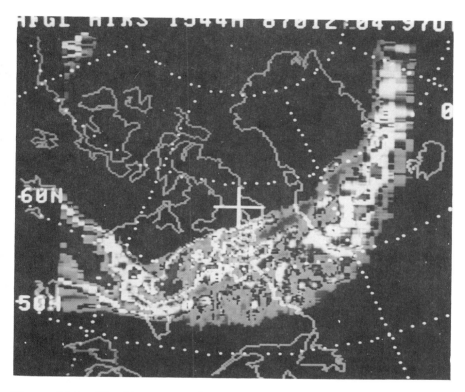

Figure 10.4: Simultaneous multispectral auroral images from the Polar BEAR satellite, N_2 LBH, 154.4 nm (*Huffman et al.*, 1989)

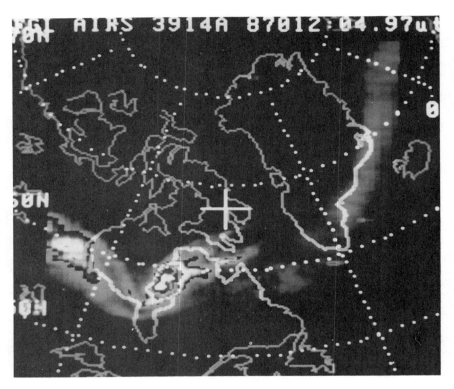

Figure 10.5: Simultaneous multispectral auroral images from the Polar BEAR satellite, N_2^+ 1N, 391.4 nm (*Huffman et al.*, 1989)

10.3. AURORAL EMISSION MODELING

by better resolution is the 215 nm "mystery" feature, which was first clearly seen in the aurora but confused with a NO γ band. It is now assigned to emission from N^+, as suggested by *Dick*, 1978.

In addition to energetic electrons, the auroral zone may have energetic ion precipitation. Protons are the most common ion, as identified through the observation of the Balmer lines of neutral atomic hydrogen in the visible region. In the ultraviolet, increases in the hydrogen Lyman alpha line at 121.6 nm have been used to characterize proton auroras, *Ishimoto et al.*, 1989a. There is some evidence for energetic oxygen ions as well, *Ishimoto et al.*, 1989b.

10.3 Auroral emission modeling

Auroral emission modeling combines what is known about the energetic particle flux, cross sections, and atmospheres to calculate the light emission from the aurora. It has been of interest for many years to also do the reverse and utilize the emission from the aurora for remote sensing of the energy and number of precipitating energetic particles. One goal would be to obtain the global energy input into the Earth's atmosphere from the energetic particles creating the aurora. This work is primarily a research effort at this time.

Modeling of auroral emission has been summarized by *Vallance Jones*, 1974. The reader should consult this reference and the other references given in this chapter for earlier work in this field. In addition to many visible region features, the near ultraviolet N_2^+ band at 391.4 nm in the first negative system has been used for many years to indicate the level of electron flux. Many characteristics of the aurora and the atmosphere can be remotely sensed from ground observations in the auroral region. For example, atmospheric temperatures can be measured from the rotational structure of the 391.4 nm and other nitrogen first negative bands, *Hunten et al.*, 1963.

In order to interpret and model the observed auroral spectrum, cross sections for the excitation of emission must be known. An account of the collisional physics involved in auroral excitation is given by *Rees*, 1989. The physics of electron collisions with atoms and molecules is treated in detail by *McDaniel*, 1989, and in other books on atomic and molecular collision physics.

Auroral emission at FUV (*Strickland et al.*, 1983) and MUV (*Daniell and Strickland*, 1986) wavelengths excited by energetic electrons has been modeled. Both Maxwellian and Gaussian electron energy distributions were used. The emission rate in Rayleighs as a function of the characteristic energy for each distribution is given for a number of prominent features such as oxygen 135.6 and 130.4 nm and nitrogen LBH bands in the FUV. In the MUV, the nitrogen Vegard Kaplan and second positive bands and the oxygen atom 297.2 nm line are modeled.

As an example of the results of such calculations, Figure 10.6 from *Strickland*, 1991, gives the intensity as a function of characteristic energy for FUV features imaged by the Polar BEAR satellite and shown in Figures 10.3, 10.4, and 10.5. Only the Maxwellian distribution is shown in the figure. Use of calculations of the type shown in Figure 10.6 together with multispectral images may allow an optical method to be developed for remotely sensing the incoming particle flux, but a large amount of additional experimental and modeling research is required.

Possibly the most extensive modeling effort of this type to date has been by *Rees et al.*, 1988. The large Dynamics Explorer data base of auroral images at ultraviolet wavelengths was used. Recent discussions of auroral excitation, emission, and associated remote sensing have been by *Rees and Lummerzheim*, 1991, and by *Meier and Strickland*, 1991.

10.4 References

Akasofu, S. I., The aurora, *Scientific American*, Dec. 1965.

Akasofu, S. I., *Polar and Magnetosphere Substorms*, D. Reidel, 1968.

Akasofu, S. I., The aurora, *American Scientist, 69*, 492–499, 1981.

Akasofu, S. I., The dynamic aurora, *Scientific American*, p. 90–97, May, 1989.

Akasofu, S. I. and S. Chapman, *Solar-Terrestial Physics*, Oxford U. Press, 1972.

Akasofu, S. I. and Y. Kamide, *The Solar Wind and the Earth*, D. Reidel, 1987.

Brekke, A. and Egeland, A., *The Northern Lights: From Mythology to Space Research*, Springer-Verlag, 1983.

10.4. REFERENCES

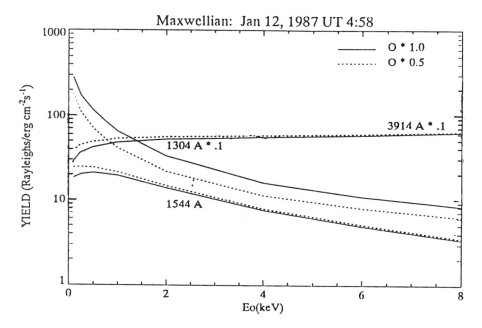

Figure 10.6: Intensity of ultraviolet emission features from precipitating energetic electrons for Polar BEAR emission features shown in Figures 10.3, 10.4, and 10.5 (*Strickland*, 1991)

Burke, W. J., D. A. Hardy, and R. P. Vancour, Chapter 8, Magnetospheric and high latitude ionospheric electrodynamics, in *Handbook of Geophysics and the Space Environment*, A. S. Jursa, editor, Air Force Geophysics Laboratory, NTIS document accession number ADA 167000, 1985.

Chamberlain, J. W., *Physics of the Aurora and Airglow*, Academic Press, 1961.

Chubb, T. A. and G. T. Hicks, Observations of the aurora in the far ultraviolet from OGO 4, *J. Geophys. Res.*, *75*, 1290–1311, 1970.

Clark, M. A. and P. H. Metzger, Auroral Lyman-Alpha observations, *J. Geophys. Res.*, *74*, 6257–6265, 1969.

Daniell, Jr., R. E. and D. J. Strickland, Dependence of auroral middle UV emissions on the incident electron spectrum and neutral atmosphere, *J. Geophys. Res.*, *91*, 321–327, 1986.

Davies, K., *Ionospheric Radio*, Peter Peregrinus Ltd., London, 1989.

Deehr, C. S. and J. A. Holtet, Editors, *Exploration of the Polar Upper Atmosphere*, Proceedings of Institute, Lillehammer, Norway, 1980, NATO Advanced Study Institute Series C, Vol. 64, Reidel, 1981.

Dick, K. A., The auroral 2150 Åfeature: A contribution from lines of singly ionized atomic nitrogen, *Geophys. Res. Lett.*, *5*, 273, 1978.

Eastes, R. W. and W. E. Sharp, Rocket-borne spectroscopic measurements in the ultraviolet aurora: The Lyman-Birge-Hopfield bands, *J. Geophys. Res.*, *92*, 10095–10100, 1987.

Eather, R. H., *Majestic Lights*, Amer. Geophys. Union, 1980.

Feldstein, Y. L. and G. V. Starkov, The auroral oval and the boundary of closed field lines of geomagnetic field, *Planet. Space Sci.*, *18*, 501, 1970.

Feldstein, Y. L. and Yu. I. Galperin, The auroral luminosity structure in the high-latitude upper atmosphere: its dynamics and relationship to the large-scale structure of the earth's magnetosphere, *Rev. Geophys.*, *23*, 217, 1985.

10.4. REFERENCES

Gerard, J. C. and C. A. Barth, OGO-4 observations of the ultraviolet auroral spectrum, *Planet. Spa. Sci., 24*, 1059–1063, 1976.

Huffman, R. E., F. J. LeBlanc, J. C. Larrabee, and D. E. Paulsen, Satellite vacuum ultraviolet airglow and auroral observations, *J. Geophys. Res., 85*, 2201–2210, 1980.

Huffman, R. E., F. P. DelGreco, R. W. Eastes, and C. I. Meng, unpublished data, 1989.

Hunten, D. M., E. G. Rawson, and J. K. Walker, Rapid measurement of N_2^+ rotational temperatures in aurora, *Canad. J. Phys., 41*, 258–270, 1963.

Ishimoto, M., C. I. Meng, G. R. Romick, and R. E. Huffman, Auroral electron energy and flux from molecular nitrogen ultraviolet emissions observed by the S3-4 satellite, *J. Geophys. Res., 93*, 9854–9866, 1988.

Ishimoto, M., C. I. Meng, G. R. Romick, and R. E. Huffman, Doppler shift of auroral Lyman α observed from a satellite, *Geophys. Res. Lett., 16*, 143–146, 1989a.

Ishimoto, M., C. I. Meng, G. R. Romick, and R. E. Huffman, Anomalous UV auroral spectra during a large magnetic disturbance, *J. Geophys. Res., 94*, 6955–6960, 1989b.

Joki, E. G. and J. E. Evans, Satellite measurements of auroral ultraviolet and 3914 Å radiation, *J. Geophys. Res., 74*, 4677–4686, 1969.

Kelley, M. C., *The Earth's Ionosphere: Plasma Physics and Electrodynamics*, Academic Press, 1989.

McCormac, B., editor, *Aurora and Airglow*, Reinhold, 1967.

McDaniel, E. W., *Atomic Collisions, Electron and Photon Projectiles*, Wiley, 1989.

Meier, R. R. and D. J. Strickland, Auroral emission processes and remote sensing, in *Auroral Physics*, C. I. Meng, M. J. Rycroft, and L. A. Frank, editors, Cambridge, 37–50, 1991.

Meng, C. I., M. J. Rycroft, and L. A. Frank, editors, *Auroral Physics*, Cambridge, 1991.

O'Neil, R. R. and R. H. Picard, Chapter 12, The Aurora, 12.2, Auroral optical and infrared emissions, in *Handbook of Geophysics and the Space Environment*, A. S. Jursa, editor, Air Force Geophysics Laboratory, NTIS document accession number ADA 167000, 1985.

Paresce, F., S. Chakrabarti, R. Kimble, and S. Bowyer, The EUV spectrum of day and nightside aurorae: 800-1400 Å, *J. Geophys. Res.*, *88*, 4905–4910, 1983.

Rees, M. H., *Physics and Chemistry of the Upper Atmosphere*, Cambridge U. Press, 1989.

Rees, M. H. and D. Lummerzheim, Auroral excitation processes, in *Auroral Physics*, C. I. Meng, M. J. Rycroft, and L. A. Frank, editors, Cambridge, 29–36, 1991.

Rees, M. H., D. Lummerzheim, R. G. Roble, J. D. Winningham, J. D. Craven, and L. A. Frank, Auroral energy deposition rate, characteristic electron energy, and ionospheric parameters derived from Dynamics Explorer 1 images, *J. Geophys. Res.*, *93*, 12,841–12,860, 1988.

Rich, F. J., Chapter 9, Ionospheric Physics, Section 9.2, High Latitude Phenomena, in *Handbook of Geophysics and the Space Environment*, A. S. Jursa, editior, Air Force Geophysics Laboratory, NTIS document accession number ADA 167000, 1985.

Strickland, D. J., private communication, 1991.

Strickland, D. J., J. R. Jasperse, and J. A. Whalen, Dependence of auroral FUV emissions on the incident electron spectrum and neutral atmosphere, *J. Geophys. Res.*, *88*, 8051–8062, 1983.

Vallance Jones, A., *Aurora*, D. Reidel, 1974.

Vallance Jones, A., Overview of auroral spectroscopy, in *Auroral Physics*, C. I. Meng, M. J. Rycroft, and L. A. Frank, editors, Cambridge, 15–28, 1991.

Whalen, J. A., Chapter 12, The Aurora, 12.1, Phenomenology, morphology, and occurrence, in *Handbook of Geophysics and the Space Environment*, A. S. Jursa, editor, Air Force Geophysics Laboratory, NTIS document accession number ADA 167000, 1985.

Chapter 11

Scattering and Fluorescence

This chapter is a brief introduction to scattering and fluorescence in the atmosphere. The scattering process results in a change of direction of the photon without a change in wavelength. Fluorescence results from excitation by photon absorption to specific upper energy levels followed by emission to lower energy levels. Scattering and fluorescence are dominant compared to airglow, discussed in Chapter 9, for many wavelengths and viewing conditions.

The chapter sections cover Rayleigh scattering in the troposphere and stratosphere; fluorescence scattering by neutral and ionized atoms and molecules, typically of more importance in the mesosphere and thermosphere; and polar mesospheric clouds. The polar mesospheric clouds, seen at ultraviolet wavelengths due to solar scatter, are an interesting recent discovery whose application to atmospheric sensing remains to be fully exploited.

11.1 Rayleigh scattering

Scattering in the atmosphere can be either from molecules, called Rayleigh scattering, or from the larger solid or liquid aerosols, called Mie scattering. Our main concern is with the Rayleigh scattering occuring at stratospheric and higher altitudes and due to the scattering of sunlight in the middle ultraviolet, (200 to 300 nm) and the near ultraviolet, (300 to 400 nm). Over much of this region, Rayleigh scattering of sunlight is the dominant radiance source when looking toward the Earth from space. This scattering radiance source is utilized in the measurement of stratospheric ozone, as discussed in more detail in Chapter 14.

For our purposes, Rayleigh scattering becomes important when the scattering cross section becomes comparable to the absorption cross section, or when the optical depth due to scattering becomes large. As discussed in Chapter 7, the Rayleigh scattering cross section is about $2 \times 10^{-25}\, cm^2$ at 300 nm and about $3 \times 10^{-22}\, cm^2$ at 50 nm. The scattering cross section varies approximately as the inverse fourth power of the wavelength. The strong absorption of ozone and molecular oxygen reduces and modifies the effects of Rayleigh scattering in the atmosphere at wavelengths shorter than about 250 nm.

At stratospheric and tropospheric altitudes, scattering is the dominant radiative transfer process, with scattering from aerosols, clouds, and larger particles becoming more important at the lower altitudes. Multiple scattering and the nonisotropic nature of the scattered radiation must be included. The radiation field active in photodissociation is dependent on ground and atmospheric scattering as well as on the direct solar flux. In extreme cases, the best image of a point source is a diffuse aureole due to multiple scattering from aerosols and dust.

Experimental measurements of the atmospheric radiance dominated by Rayleigh scattering were an early goal of sounding rocket and satellite experimentation. These measurements were driven largely by the objective of developing a remote sensing method for ozone based on the absorption of the scattered solar emission by stratospheric ozone. These measurements were reviewed by *Rawcliffe and Elliot*, 1966. The large programs for ozone measurement from satellites using the backscattered UV approach now in place, such as TOMS, routinely make radiance measurements at the specific wavelengths of interest for ozone monitoring. These data bases represent additional measurements of the Rayleigh scatter background which may be examined for previously overlooked geophysical effects. Ozone measurements from satellites are described in Chapter 14.

Exploratory satellite experiments with the Rayleigh scatter region within their spectral range include OGO-4: *Barth and Mackey*, 1969; *Anderson*, 1969; and S3-4: *Huffman et al.*, 1980; *Huffman et al.*, 1989. The data bases generated by these programs have been influential in the design of specific remote sensing investigations and in the development of radiance codes.

Figure 11.1 illustrates data from the S3-4 satellite as well as from several earlier measurements. The points shown in the figure are from *Rawcliffe and Elliot*, 1966; *Elliot et al.*, 1967, and *Barth and Mackey*, 1969. Note that the nadir background is structured due to the Franhofer absorption structure

11.1. RAYLEIGH SCATTERING

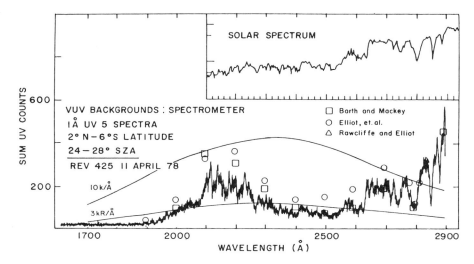

Figure 11.1: MUV nadir atmospheric radiance from the S3-4 satellite (*Huffman et al.*, 1980)

in the solar flux. A solar flux measurement at approximately the resolution of the atmospheric radiance measurement due to *Broadfoot*, 1972, is shown in the inset to Figure 11.1. The correlation of solar flux and atmospheric radiance structure is evident. Measurements of this type compare well with the LOWTRAN 7 code, as shown in Chapter 13.

Sounding rocket measurements have continued to provide important information about the Rayleigh scattering radiance in the atmosphere, but they are not reviewed in detail here. Measurements associated with missile defense applications are enabling tests of the models over a range of solar zenith angles, scattering angles, and limb observation directions. These measurements or measurement plans are described by *Carbary and Meng*, 1989; *Strickland et al.*, 1989; *Smathers et al.*, 1989; and *Bechis*, 1989.

Theoretical calculation of the radiance of the atmosphere in the middle ultraviolet is discussed by *Green*, 1966, and *Dowling and Green*, 1966, who review the situation regarding calculations and experiments up to that time. Improvements in our knowledge of the atmosphere and an understanding of the needs for radiance and transmission calculations has led to improved models.

Recent model calculations by *Strickland et al.*, 1988, include the angular

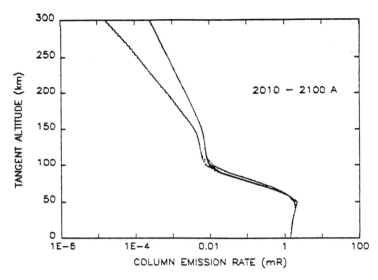

Figure 11.2: MUV limb radiance profile, 201–210 nm (*Strickland et al.*, 1988)

dependence of the scattering and the contribution from NO fluorescence. A set of calculations for a SZA of $30°$ is shown in Figures 11.2, 11.3, 11.4, and 11.5.

Scattering contributions to radiance are included in a convenient way for the nonspecialist in transmission and radiance codes such as LOWTRAN 7, as covered in Chapter 13. The current lower limit of 200 nm covers the primary region of UV radiance influenced by scattering. The codes are designed for use at tropospheric altitudes as well, and so they include aerosol models and multiple scattering (*Isaacs et al.*, 1987). Codes such as LOWTRAN 7, MODTRAN, and AURIC are the preferred approach for many applications.

To fully test the models, measurements specifically addressing the scattering radiance problem will be required. Improved measurements are needed to overcome interference from out-of-band emission in the measurements. In addition, observations near the solar direction and under twilight conditions are needed.

Because of the intensity of the scatter radiance and its low altitude of origin, it is likely that additional remote sensing applications will be developed utilizing, or requiring knowledge of, the scatter radiance. These

11.1. RAYLEIGH SCATTERING

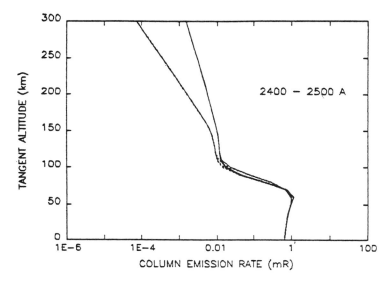

Figure 11.3: MUV limb radiance profile, 240–250 nm (*Strickland et al.*, 1988)

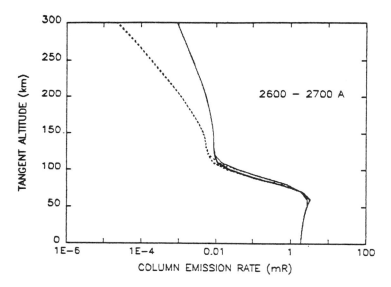

Figure 11.4: MUV limb radiance profile, 260–270 nm (*Strickland et al.*, 1988)

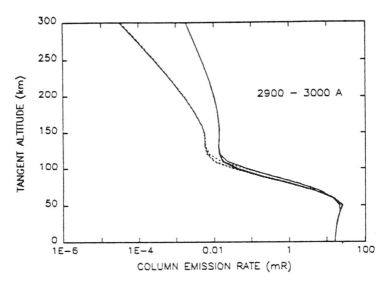

Figure 11.5: MUV limb radiance profile, 290–300 nm (*Strickland et al.*, 1988)

might involve geophysical phenomena such as **volcano emission tracking** and **gravity waves**. General applications such as **stratospheric aircraft traffic control** may be possible. These phenomena have been seen at times from ozone backscatter sensors or other atmospheric measurements.

11.2 Fluorescence

Fluorescence is the prompt emission of absorbed photons. It occurs at specific wavelengths corresponding to differences in energy levels of the atomic or molecular emitter. In this chapter, the term fluorescence also includes resonance scattering. In resonance scattering, the emitted photon is at the same nominal wavelength as the absorbed photon. In the atmosphere, this generally leads to multiple scattering. Multiple scattering is a major determinant of the emission observed. The cross section for reabsorption can be very high, resulting in what is called optically thick conditions.

The most intense fluorescence and the greatest amount of multiple scattering for atoms occurs at the *resonance lines* between ground electronic states and the lowest energy electronic state that can undergo a transition that is allowed by spectroscopic selection rules. In the atmosphere, examples

11.2. FLUORESCENCE

are the atomic hydrogen Lyman α line at 121.6 nm and the atomic oxygen resonance multiplet consisting of several lines centered at 130.4 nm.

At each cycle of absorption and reemission, the frequency of the photon may change slightly, and thus some photons eventually are shifted from the most intense absorption frequency at the center of the line. In this way the "trapped" photons eventually "escape" in the wings of the line, where the absorption cross section in smaller. Multiple scattering of resonance lines is a complicated subject with many aspects of the problem difficult to include in models. The review by *Meier*, 1991, is especially valuable for more details on scattering at intense lines.

Fluorescence is considered apart from airglow in this book to stress the difference in the emission processes involved. The term airglow is reserved for emission arising from chemical and lower energy collision processes generally occurring outside the auroral regions, as discussed in Chapter 9. In the atmosphere, both fluorescence and airglow may contribute to the emission for any given feature. The relative importance of the two is usually clear, but it may change with geophysical and viewing conditions, as will be indicated for the oxygen atom resonance multiplet near 130.4 nm.

The fluorescence sources in the atmosphere, considered as a whole, are introduced here. The emissions that form the basis for more widely used remote sensing applications are discussed individually in the appropriate places in Chapters 15 and 17.

Atmospheric emissions in the UV due to fluorescence are listed below in order of decreasing wavelength:

N_2^+, **1st negative bands** These bands are relatively strong when seen in the limb or against a background that does not include the strong scatter in the same region. The (0,0) band at 391.4 nm is just within our definition of the UV range, but for ground observation the nearby 425.8 nm (1,0) band is preferred because it is not subject to the self absorption found for the ground state. The bands are also excited by high energy particles in the aurora, as discussed in the last chapter.

OH, hydroxyl radical This fluorescence has been observed in the Earth limb with high resolution UV spectroscopy. It is a minor constituent of the 40 to 80 km altitude region of the mesosphere, but like NO, which also primarily occurs in the mesosphere, it is a participant in many reactions and its concentration is important in modeling. The primary emission is the (0,0) band of the $A\,^2\Sigma - X\,^2\Pi$ band system, with the strongest lines near 308 nm. A review and description of the planned MAHRS sensor for OH and NO measurements from the shuttle is given by *Conway et al.*, 1988. The sensor will have a resolution of 0.01 nm, which is one of the highest resolutions of any planned experiment. Some previous observations of OH fluorescence have been with a spectrometer on a sounding rocket, *Anderson*, 1971a,b, and with a spectrometer on a balloon, *Torr et al.*, 1987. Other OH measurements are given in Chapter 14.

Mg^+, **magnesium ion** Scattering from the magnesium ion near 280 nm is an ultraviolet representative of similar scattering in the visible region from sodium and other trace metals and metal ions found in the upper atmosphere. See *Gérard et al.*, 1979, for observations and *Hays et al.*, 1973, for a description of the Visible Airglow Experiment on Atmospheric Explorer E that was used to obtain these measurements.

NO, gamma bands These bands, observed in the limb, form the basis for a remote sensing method for NO, as discussed in detail in Chapter 15. There are several bands in the 210–250 nm region.

O, atomic oxygen The oxygen resonance multiplet centered near 130.4 nm can be the brightest contributor to the UV radiance from the Earth in the FUV, with an intensity of 10 to 15 kiloRayleighs. This emission is very complicated, however, being composed generally of multiply scattered sunlight. There are situations, however, when the photoelectron excitation mechanism becomes dominant. The scattering problems are so formidable that this line is usually not considered for remote sensing, as discussed further in Chapter 15. The nearby atomic oxygen multiplet usually called the 135.6 nm line is used instead to monitor atomic oxygen. A limb scan is shown in Figure 11.6, where the difference between 130.4 and 135.6 nm is readily apparent. Also shown is the limb scan of NI 149.6 nm, which is produced entirely from airglow.

H, Lyman α The resonance fluorescence scattering of the intense solar line at 121.6 nm by the trace of atomic hydrogen in the atmosphere leads to the well-known hydrogen *geocorona* of the Earth. Depending on geophysical conditions, this line can be from 10 to 20 kiloRayleighs maximum in the daytime and about one-tenth of these values at night. It is discussed as the basis of a remote sensing method in Chapter 15.

The second member of the resonance series, Lyman β at 102.5 nm, is also an important radiance source.

O^+, **atomic oxygen ion** The atomic oxygen ion multiplet near 83.4 nm is a strong contributor to the atmospheric radiance in the longer wavelength part of the EUV. The initial emission of the line is due to solar photoionization, but the observed radiance in the Earth limb is due to scattering of the initial emission by atomic oxygen ions centered near 250 km in the F region of the ionosphere. It has been proposed as the basis for an ionospheric electron density measurement technique, as discussed in Chapter 17.

He, helium The 58.4 nm resonance line has been proposed as a remote sensing method for thermospheric helium, as discussed in Chapter 15.

He^+, **helium ion** The 30.4 resonance line of singly ionized helium is prominent in this part of the EUV region, when the Earth's atmosphere is observed from space. **Magnetospheric imaging** using solar fluorescence from He^+ and O^+ is discussed in the global imaging chapter, Section 16.3.

Weak fluorescence sources in the radiance from the Earth's atmosphere have been sought for use in remote sensing. The search for molecular hydrogen emission excited by solar Lyman β was reported by *Feldman and Takacs*, 1975. While the hydrogen emission could not be found in the spectra, consistent with model predictions, improved sensors may allow this or similar approaches to be developed.

Multiple self-scattering of molecular emission bands is generally not an effect found in the atmosphere. However, for the bands of the abundant molecular nitrogen, some consideration of this effect may be necessary. The nitrogen band scattering in the Lyman-Birge-Hopfield bands has been discussed by *Conway*, 1982, 1983.

11.3 Polar mesospheric clouds

Polar mesospheric clouds, or PMC, have been found in satellite UV observations of the mesosphere in the limb direction with the Solar Mesospheric Explorer, (SME), *Thomas*, 1984. These emission layers are now believed to be more or less identical with the long known **noctilucent clouds**, or NLC, which have been studied by ground observers in polar regions for many years.

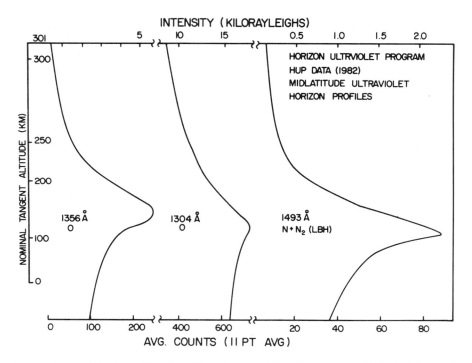

Figure 11.6: Limb profiles for OI 130.4 nm, OI 135.6 nm, and NI/LBH 149.3 nm from HUP shuttle flight (day, midlatitude, near overhead sun)

11.3. POLAR MESOSPHERIC CLOUDS

There may be subtle differences still unrecognized, and the term PMC will be used here, as we are primarily concerned with their use as a UV remote sensing method or as a possibly unrecognized additional component of the background. The PMC are seen because of scattering by the particles of the cloud in the near and mid UV.

PMCs are found at latitudes of $50°$ and higher at both poles. They occur as thin layers at about 85 km, which is close to or at the mesopause. This temperature reaches a seasonal minimum in the summer, when the clouds are brightest. Indeed, the appearance and brightness of PMC are well correlated with this seasonal minimum mesopause temperature. There is also evidence for a solar cycle variation. The clouds are only found about four weeks before and eleven weeks after the summer solstice. Differences between PMC and NLC are believed to be due to differences in observation location. More details about both are in *Thomas and Olivero*, 1989, and in a recent book on NLC by *Gadsden and Schroder*, 1989. These references will also direct the interested reader to further publications on PMC.

Figure 11.7 illustrates the limb profile changes found when PMC are present. It is based on data in *Thomas*, 1984. The intensities of the PMC may be 10 to 100 times the local Rayleigh scatter background. These large increases are due to the so-called amplification factor caused by limb viewing of an altitude layered radiance source such as a cloud or airglow layer.

For our applications, polar mesospheric clouds are similar to the aurora in that they are a discrete radiance source that may be either a problem with other observations or an interesting phenomenon for study in itself. PMC have now also been observed in the nadir direction, but they are difficult to find, *Thomas et al.*, 1991. There is a recent review, *Thomas*, 1991.

The **Solar Mesosphere Explorer** (SME) satellite had as its main research objectives the measurement of ozone, nitric oxide, and solar flux by using UV and IR measurements from a spinning satellite. As it relies on solar scatter for atmospheric sensing, it will be described in some detail as a prototype of possible atmospheric monitoring methods in future programs. The ozone and nitric oxide measurements are mentioned further in Chapters 14 and 15, respectively.

The SME was in a sun-synchronous orbit at about 0300–1500 local time for equator crossings and at an altitude of 534 km. The limb spatial resolution was about 3.5 km at a range of 2700 km. It was launched October 6, 1981, and had a lifetime covering a number of years. There is an extensive data base.

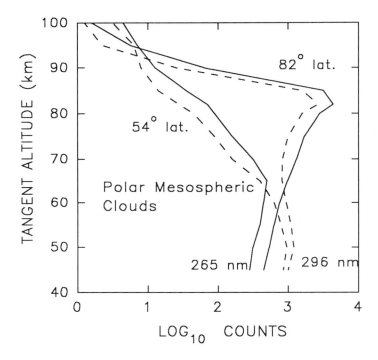

Figure 11.7: Effect of polar mesospheric clouds on MUV limb profile measured by the SME satellite (based on data in *Thomas*, 1984)

(Reprinted with permission from Pergamon Press, Ltd.)

UV atmospheric measurements were made with a one-eighth meter Ebert spectrometer with a wavelength range of 200 to 330 nm and a spectral resolution of 1.5 nm. There were two EMR type F photomultiplier detectors separated by about 30 nm. For ozone, the wavelengths of 265 and 297 nm were used, and for nitric oxide, the gamma band at about 215 nm was the principal data source. Launch sensitivities were 0.11, 0.12, and 0.03 counts per Rayleigh-second at 220, 260, and 300 nm, respectively. The solar flux measurement range was 120 to 310 nm, with the H Lyman α line at 121.6 nm a primary measurement objective.

The SME principal investigator is C. A. Barth of the Laboratory for Atmospheric and Space Physics, University of Colorado, and the satellite program was sponsored by NASA. A description of the program is given by *Thomas et al.*, 1980, and the initial results are described in *Barth et al.*, 1983.

Polar mesospheric clouds, and UV remote sensing in general, may turn out to be an important monitor of **global change**. There seems to be little or no evidence of the occurrence of NLC, and thus apparently PMC, before the large influx of pollutant gases that began about the beginning of the twentieth century. In particular, the increase in methane in the atmosphere, leading to an increase in water vapor at mesospheric altitudes, has been suggested as the reason for the appearance of the high altitude clouds (see *Thomas et al.*, 1989, and references therein). If further investigation confirms a relationship between PMC and atmospheric methane, it will mean that PMC observations should be considered for use as a monitor of global change.

11.4 References

Anderson, G. P., C. A. Barth, F. Cayla, and J. London, Satellite observations of the vertical ozone distribution in the upper stratosphere, *Ann. Geophys.*, *25*, 341, 1969.

Anderson, J. G., Rocket-borne ultraviolet spectrometer measurement of OH fluorescence with a diffusive transport model for mesospheric photochemistry, *J. Geophys. Res.*, *76*, 4634, 1971a.

Anderson, J. G., Rocket measurement of OH in the mesosphere, *J. Geophys. Res.*, *76*, 7820–7824, 1971b.

CHAPTER 11. SCATTERING AND FLUORESCENCE

Barth, C. A. and E. F. Mackey, OGO-4 ultraviolet airglow spectrometer, *IEEE Trans. Geosci. Electron., GE-7*, 114, 1969.

Barth, C. A., D. W. Rusch, R. J. Thomas, G. M. Mount, G. J. Rottman, G. E. Thomas, R. W. Sanders, and G. M. Lawrence, Solar Mesospheric Explorer: Scientific objectives and results, *Geophysical Research Letters, 10*, 237–240, 1983.

Bechis, K. P., UV observations during the STARLAB space shuttle mission, *Ultraviolet Technology III*, R. E. Huffman, editor, *Proc. SPIE, 1158*, 232–241, 1989.

Broadfoot, A. L., The solar spectrum 2100–3200 Å, *Astrophys. J., 173*, 681–689, 1972.

Carbary, J. F. and C. I. Meng, Limb profiles from low earth orbit, *Ultraviolet Technology III*, R. E. Huffman, editor, *Proc. SPIE, 1158*, 51–58, 1989.

Conway, R. R., Self absorption of the N_2 Lyman-Birge-Hopfield bands in the far ultraviolet dayglow, *J. Geophys. Res., 87*, 859, 1982.

Conway, R. R., Multiple fluorescent scattering of N_2 ultraviolet emissions in the atmospheres of the Earth and Titan, *J. Geophys. Res., 88*, 4784, 1983.

Conway, R. R., D. K. Prinz, and G. H. Mount, Middle atmosphere high resolution spectrograph, *Ultraviolet Technology II*, R. E. Huffman, editor, *Proc. SPIE, 932*, 50–60, 1988.

Dowling, J. and A. E. S. Green, Second-order scattering contributions to the Earth's radiance in the middle ultraviolet, Chapter 6 in *The Middle Ultraviolet: Its Science and Technology*, A. E. S. Green, Editor, Wiley, 1966.

Elliot, D. D., M. A. Clark, and R. D. Hudson, *Latitude Distribution of the Daytime Ozone Profile above the Peak*, Air Force Report No. SAMSO-TR-67-39, Aerospace Report No. TR-0158 (3260-10)-2, September, 1967.

Feldman, P. D. and P. Z. Takacs, A search for molecular hydrogen fluorescence near 100 km, *J. Atmos. Sci., 32*, 2209–2212, 1975.

11.4. REFERENCES

Gadsden, M. and W. Schroder, *Noctilucent Clouds*, Springer-Verlag, 1989.

Gérard, J. C., D. W. Rusch, P. B. Hays, and C. L. Fesen, The morphology of equatorial Mg^+ ion distribution deduced from 2800 Å airglow observations, *J. Geophys. Res., 84*, 5249–5258, 1979.

Green, A. E. S., The radiance of the Earth in the middle ultraviolet, Chapter 5 in *The Middle Ultraviolet: Its Science and Technology*, A. E. S. Green, Editor, Wiley, 1966.

Hays, P. B., G. Carignan, B. C. Kennedy, G. G. Shepard, and J. G. Walker, The Visible Airglow Experiment on Atmospheric Explorer, *Radio Sci., 8*, 369, 1973.

Huffman, R. E., F. J. LeBlanc, J. C. Larrabee, and D. E. Paulsen, Satellite vacuum ultraviolet airglow and auroral observations, *J. Geophys. Res., 85*, 2201–2215, 1980.

Huffman, R. E., L. A. Hall, and F. J. LeBlanc, Comparison of the LOWTRAN 7 code and S3-4 satellite measurements of UV radiance, *Ultraviolet Technology III*, R. E. Huffman, editor, *Proc. SPIE, 1158*, 38–44, 1989.

Isaacs, R. G., W.-C. Wang, R. D. Worsham, and S. Goldenberg, Multiple scattering LOWTRAN and FASCODE models, *Appl. Opt., 26*, 1272–1281, 1987.

Meier, R. R., Ultraviolet spectroscopy and remote sensing of the upper atmosphere, *Space Sci. Rev., 58*, 1–135, 1991.

Rawcliffe, R. D., and D. D. Elliot, Latitude distribution of ozone at high altitudes deduced from satellite measurements of the earth: Radiance at 8240 Å, *J. Geophys. Res., 71*, 5077, 1966.

Smathers, H. W., G. R. Carruthers, W. D. Ramsey, G. Steiner, W. Louissaint, Calibration and performance of the Ultraviolet Plume Instrument, *Ultraviolet Technology III*, R. E. Huffman, editor, *Proc. SPIE, 1158*, 212–231, 1989.

Strickland, D. J., R. P. Barnes, and D. E. Anderson, Jr., *UV background calculations: Rayleigh scattered and dayglow backgrounds from 1200 to 3000 Å*, AFGL-TR-88-0200, Air Force Geophysics Laboratory, 1988.

Strickland, D. J., R. P. Barnes, R. J. Cox, D. E. Anderson, J. F. Carbary, and C. I. Meng, Analysis of UV limb data from low earth orbit, *Ultraviolet Technology III*, R. E. Huffman, editor, *Proc. SPIE, II58*, 59–68, 1989.

Thomas, G. E., Solar Mesospheric Explorer measurements of polar mesospheric clouds (noctilucent clouds), *J. Atmos. Terr. Phys., 46*, 819, 1984.

Thomas, G. E., Mesospheric clouds and the physics of the mesopause region, *Rev. Geophys., 29*, 553–575, 1991.

Thomas, G. E. and J. J. Olivero, Climatology of polar mesospheric clouds, 2, Further analysis of Solar Mesospheric Explorer data, *J. Geophys. Res., 94*, 14,673–14,681, 1989.

Thomas, G. E., C. A. Barth, E. R. Hansen, C. W. Hord, G. M. Lawrence, G. H. Mount, G. J. Rottman, D. W. Rusch, A. I. Stewart, R. J. Thomas, J. London, P. L. Bailey, P. J. Crutzen, R. E. Dickinson, J. C. Gille, S. C. Liu, J. F. Noxon, and C. B. Farmer, Scientific objectives of the Solar Mesospheric Explorer mission, *Pure appl. Geophys., 118*, 591–615, 1980.

Thomas, G. E., J. J. Olivero, E. J. Jensen, W. Schroeder, and O. B. Toon, Relation between increasing methane and the presence of ice clouds at the mesopause, *Nature, 338*, 490–492, 1989.

Thomas, G. E., R. D. McPeters, and E. J. Jensen, Satellite observations of polar mesospheric clouds by the Solar Backscattered Ultraviolet spectral radiometer: Evidence of a solar cycle dependence, *J. Geophys. Res., 96*, 927–939, 1991.

Torr, D. G., M. R. Torr, W. Swift, J. Fennelly, and G. Liu, Measurement of OH (X $^2\Pi$) in the stratosphere by high resolution UV spectroscopy, *Geophys. Res. Lett., 14*, 937–940, 1987.

Chapter 12

Atmospheric Ultraviolet Backgrounds

This chapter discusses the atmospheric ultraviolet background, which is the term used here to represent all the naturally occurring radiance sources in the atmosphere. The atmospheric background thus is primarily due to the airglow, aurora, and scattering sources of radiance covered in the last three chapters. The background differs for auroral and nonauroral locations and from day to night. Background levels for these cases are given in this chapter, together with a discussion of variability due to clutter.

Although most of the background radiance levels given in this chapter are the result of measurement, in the future the atmospheric radiance will be more readily obtained from UV radiance and transmission codes such as AURIC (Atmospheric Ultraviolet Radiance Integrated Code), as discussed in the next chapter. Indeed, many UV background problems can be successfully addressed today with LOWTRAN 7. Recent theoretical modeling is discussed by *Anderson and Strickland*, 1989.

This chapter does not discuss **celestial sources** of the ultraviolet, including stars, planets, comets, and possible scattering from dust. UV stars are briefly discussed in Chapter 5, in connection with their use in calibration. A recent review of many of the celestial sources, and of the effects of atmospheric radiance on astronomical observations, is by *Henry*, 1991. To date, celestial sources have been largely ignored as backgrounds in atmospheric remote sensing, but this situation will change in the future.

The **background** depends on the objective of a given measurement or application. The backgrounds concept is useful, however, if for no other reason than to emphasize the point that observation through the atmosphere, such as of stars or missile exhaust plumes, have to be made with an understanding of all the natural atmospheric radiance sources and of possible limitations due to atmospheric transmission. Any unwanted source of emission in the sensor FOV is a part of the background.

The atmospheric radiance sources discussed in this chapter, as well as UV stars and other celestial radiance sources, can both be unwanted backgrounds or desired signals depending on circumstances. The aurora, for example, may represent unwanted background interference for astronomy experiments. However, if the goal were to monitor the location of the auroral oval, the auroral emission would be the "signal" and local airglow and scattering sources, plus possibly UV stars, would be the "background". The situation is well-described by a maxim useful to keep in mind in remote sensing:

one person's signal is another person's background.

12.1 Day backgrounds

The day UV background in the 110 to 310 nm region is given in the upper part of Figure 12.1, which is an improved version of the curve in the *Handbook of Geophysics and the Space Environment, Huffman,* 1985. Measurements shown are from a number of sources.

The intention in Figure 12.1 is to give introductory information about the background by showing what are roughly the maximum and the minimum radiance levels that will be encountered. The upper curve for the day maximum is the level found for a nadir, or Earth center, view with the sun in the zenith, or directly overhead, position. The lower part of the figure gives the night minimum conditions, which are found at midlatitudes, away from both the equatorial UV airglow from the ionosphere, found in the early evening hours, and the auroral zone.

In Figure 12.1, the day radiance level is generally determined by the incident solar flux, the backscattered radiation, absorption of the backscatter by atmospheric constituents, and day airglow emissions. The most intense FUV emissions by far are the hydrogen Lyman α line at 121.6 nm and the oxygen resonance multiplet centered at 130.4 nm. Either one may be dominant in a given observation.

Rayleigh scattering alone would lead to an increase of the background with a decrease in the wavelength, as discussed in the last chapter. However,

12.1. DAY BACKGROUNDS

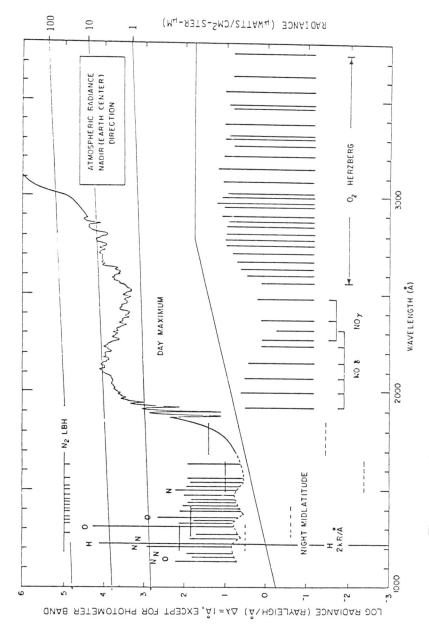

Figure 12.1: Atmospheric UV backgrounds for day and night nadir viewing in the 110 to 380 nm region. See text. (*Huffman*, 1985)

there are other effects to consider. The broad minimum in the curve near 250 nm is due to absorption by stratospheric ozone. The decrease below about 200 nm is due to absorption by oxygen in the Schumann-Runge bands and continuum. The Lyman-Birge-Hopfield bands of nitrogen become observable because oxygen absorption, and a decrease in the level of the incident solar flux toward shorter wavelengths, decreases other emission sources.

Although all readily available measurements have been taken into account in Figure 12.1, the levels shown are based primarily on the S3-4 satellite measurements from the experiment VUV Backgrounds taken in April and May, 1978, when the solar activity was relatively high and rising toward the maximum, *Huffman, et al.*, 1980. The daytime curve will rise and fall with the solar cycle, as the regions dominated by solar scattering, between roughly 200 and 300 nm, and the FUV emissions at H Lyman α, O 130.4 and 135.6 nm, and the N_2 Lyman-Birge-Hopfield (LBH) bands, are all excited directly or indirectly by the solar flux. Variability due to solar flares has been reported by *Opal*, 1973.

It must be kept in mind that emissions resulting from photoelectron excitation, such as the O 135.6 nm and the LBH band emissions are affected by the solar variability of the EUV radiation primarily responsible for their excitation rather than by the solar flux variability in the region where they emit. As we have mentioned in Chapter 7, the variability at EUV wavelengths is much larger than at longer wavelengths.

The maximum daytime level shown in Figure 12.1 is for nadir viewing, but much of the time remote sensing and other applications will be viewing at a slant path. For Earth limb directions, there is generally **limb brightening**, or a maximum radiance observed at a certain tangent altitude. This maximum is the result of low radiances due to the absorption of the radiation at lower tangent altitudes and by the gradual decrease in emission as the tangent altitude increases, due to the decrease in atmospheric density. At the tangent altitude of the peak of the limb radiance profile, the radiance levels may be 10 or 20 times the values shown, depending on the spatial and spectral resolution. Examples of limb radiance brightening are given throughout this book. It also applies to the weak night airglow emission shown in the bottom part of Figure 12.1. Viewing direction definitions are in Chapter 5.

The use of the Figure 12.1 daytime curve as a maximum radiance level must be modified for limb viewing directions. However, this increase is offset by the fact that the curve is for overhead, or $SZA = 0$ conditions, which is usually not the case. As the sun angle becomes larger, the radiance decreases,

12.1. DAY BACKGROUNDS

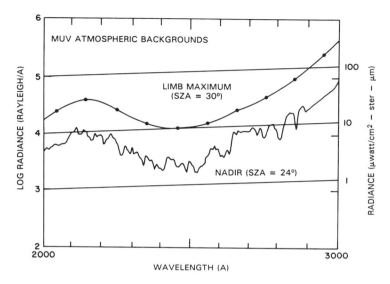

Figure 12.2: Maximum limb brightness in MUV, *Strickland, et al.*, 1988, compared with nadir values from Figure 12.1

as it is excited by solar radiation that has traversed a longer slant path and thus has been attenuated by absorption and scattering before reaching the column under observation. Thus, the noon nadir radiance is a useful background level for initial calculations of daytime conditions.

With these caveats understood, Figure 12.1 has proved to be valuable as an introduction to the radiative environment of the Earth outside the auroral region. Even for the aurora, as will be discussed, the maximum radiance level is less than the day maximum level shown in Figure 12.1. The maximum day background radiance in the MUV is shown in Figure 12.2, which is based on the scattering and fluorescence limb profiles discussed in Chapter 11, *Strickland et al.*, 1988.

Radiance units shown in Figures 12.1 and 12.2 are Rayleighs per Å for a 1 Å bandwidth in regions of continuous emission. For the discrete lines and bands, a vertical line is shown for the feature, with the height based on the assumption of a 1 Å bandwidth. Thus, the number of Rayleighs in each feature can be directly read from the figure. This approach also allows a better representation of the discrete structure on the scale of the figure. Conventional spectral plots for the major emitters are shown elsewhere in this and other chapters. The spectral resolution available throughout most of this range is to about 5 Å bandwidth (FWHM). Radiance is also given in units of $\mu watt/(cm^2)(\mu meter)$ in Figures 12.1 and 12.2.

In Figure 12.1, four day and four night intensity levels are shown in the FUV. These are the day radiance maximum values and the night upper limits from the S3-4 photometer measurements, *Huffman et al.*, 1980. This sensor is discussed at length in Chapter 4. The horizontal lengths of the lines are the full width at half maximum (FWHM) of the photometer filters. The day maximum for the band centered at 1500 Å is the lowest background level for nadir viewing in the FUV.

At the time of the S3-4 flight, there was considerable uncertainty in the intensity value of the 150 nm minimum in the day background. The OGO-4 ion chambers had measured a relatively low value, but measurements with filter photometers on the Orbiting Astronomical Observatory A-2 indicated a higher value, *Fishburne et al.*, 1972. The S3-4 spectrometer measurements, shown in Figure 12.1, are similar in radiance level to the broad band ion chamber measurements from OGO-4, and to subsequent measurements by the UV imagers on HILAT and Polar BEAR (see Chapter 16). Measurement of the radiance at 192 nm with OAO were similar to previous and subsequent measurements, and the higher value at 150 nm is possibly due to a decrease in sensitivity of this photometer in flight.

The spectral lines and bands shown as vertical lines in the day background portion of Figure 12.1 are given in Table 12.1. This table includes spectroscopic identification and observed radiance from the S3-4 satellite. These lines and bands are due to airglow and to solar scattering, as discussed in Chapters 9 and 11.

Near ultraviolet backgrounds, from 400 to 300 nm, are covered in part in Figure 12.1. However, when solar radiation can penetrate to the surface, as is the case at wavelengths longer than about 320 nm, the surface albedo must be considered. Scattering dominates the background, obscuring airglow and fluorescent emission which, however, may be seen in limb viewing measurements.

The only published measurements of the NUV nadir viewing backgrounds known to the author are due to *McPeters*, 1989. The measurements were obtained with the SBUV sensor on Nimbus 7 and are for a SZA of 80°. For planning purposes, Figure 12.3 is provided, which is a LOWTRAN 7 radiance curve using an Earth albedo of 0.2 and a near nadir SZA, *Huffman et al.*, 1989. With this readily available code and a personal computer, the interested reader can explore many other cases, as discussed in the next chapter on UV codes.

12.1. DAY BACKGROUNDS

Table 12.1: MUV and FUV day nadir emission features. See text and Figure 12.1.

Transition	λ (nm)	B (kR)	Comment
N_2 LBH (1,4)	163.-	0.09	LBH: $a^1\Pi_g - X^1\Sigma_g^+$
N_2 LBH (0,2)	155.4	0.09	
N_2 LBH (2,3)	153.-	0.07	
N_2 LBH (4,4)	150.8	0.09	
N I	149.3	0.18	$2p^3\ ^2D^o - 3s\ ^2P$
N_2 LBH (1,1)	146.4	0.10	
N_2 LBH (0,0)	145.-	0.07	also LBH (3,2)
N_2 LBH (2,1)	143.0	0.08	
N_2 LBH (1,0)	141.6	0.10	also LBH (4,2)
N_2 LBH (3,1)	139.7	0.07	also LBH (6,3)
N_2 LBH (2,0)	138.4	0.12	also LBH (5,2)
O I	135.6	0.7	$2p^4\ ^3P^o - 3s\ ^5S$
N_2 LBH (3,0)	135.4	0.1	est., blend, O I
N_2 LBH (5,1)	133.9	0.06	
N_2 LBH (4,0)	132.5	0.16	
O I	130.4	20.	$2p^4\ ^3P^o - 3s\ ^3S$
N_2 LBH (5,0)	129.8	.	blend O I
N_2 LBH (6,0)	127.3	0.06	
N I	124.3	0.08	$2p^3\ ^2D^o - 3s'\ ^2D$
H Lyman α	121.6	15.	$1s\ ^2S^o - 2p\ ^2P$
N I	120.0	0.8	$2p^3\ ^4S^o - 3s\ ^4P$
N I	116.8	0.09	$2p^2\ ^2D^o - 3d\ ^2F$
O I	115.2	0.07	$2p^4\ ^1D - 3s'\ ^1D^o$
N I	113.4	0.15	$2p^3\ ^4S^o - 2p^4\ ^4P$

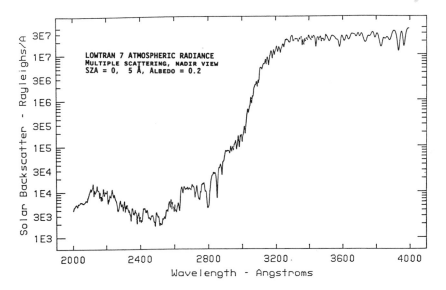

Figure 12.3: Daytime nadir background radiance, 200 to 400 nm, from LOWTRAN 7 model

Extreme ultraviolet day backgrounds have been measured from the P78-1 spacecraft, *Chakrabarti et al.*, 1983. The spectrum consists primarily of discrete lines and multiplets from atomic species. The emission observed in the nadir direction for day midlatitude from the satellite altitude of 1000 km is given in Table 12.2.

Full orbit radiance plots obtained from the S3-4 satellite for the FUV photometer bands shown in Figure 12.1 are displayed in Figures 12.4, 12.5, 12.6, and 12.7, *Huffman et al.*, 1980. These typical records for this polar orbiting satellite cover a wide range of solar zenith angles for nadir viewing. They illustrate the variation in intensity from day to night and the relationship of the auroral intensity relative to local airglow. Note the appearance of the auroral emission four times in each curve and the appearance of the equatorial airglow belts in the 134 nm filter data.

Twilight conditions are a special problem. The line of sight of a sensor will encounter a range of solar illumination conditions that are difficult to model. One interesting case is to observe a sunlit atmosphere against a dark background, as happens briefly in many satellite observations.

12.1. DAY BACKGROUNDS

Table 12.2: EUV day nadir midlatitude background emission features. See text.

Transition	λ (nm)	B (R)	Comment
N II	108.5	500	$2p^2\ ^3P - 2p^3\ ^3D^o$
O I	102.7	235	$2p^4\ ^3P - 3d\ ^3D^o$
N$_2$ BH (1,1)	100.9	150	$b^1\Pi_u - X^1\Sigma_g^+$
O I	98.9	850	$2p^4\ ^3P - 3s'\ ^3D^o$
N I	95.3	125	$2p^3\ ^4S^o - 3d\ ^4P$
N II	91.6	185	$2p^2\ ^3P - 2p^3\ ^3P^o$
O I	91.1	90	Recomb. cont.
O II	83.4	500	$2p^3\ ^4S^o - 2p^4\ ^4P$
He I	58.4	330	$1s^2\ ^1S - 2p\ ^1P$
O II	53.8	115	$2p^3\ ^4S^o - 3s\ ^4P$

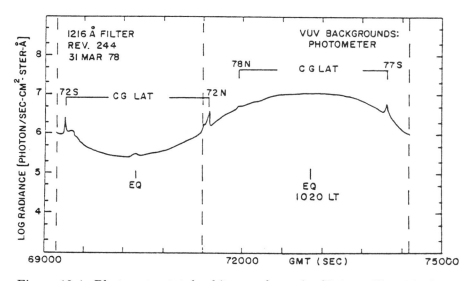

Figure 12.4: Photometer total orbit scan from the S3-4 satellite, 121.6 nm filter

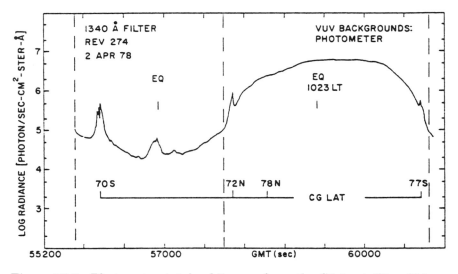

Figure 12.5: Photometer total orbit scan from the S3-4 satellite, 134 nm filter

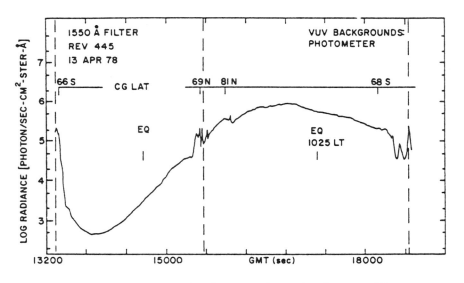

Figure 12.6: Photometer total orbit scan from the S3-4 satellite, 155 nm filter

12.2. NIGHT BACKGROUNDS

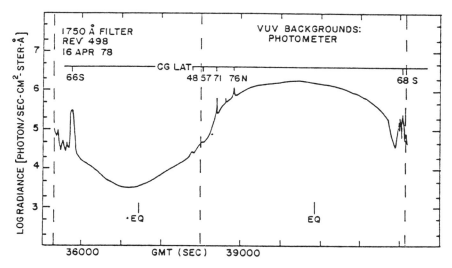

Figure 12.7: Photometer total orbit scan from the S3-4 satellite, 175 nm filter

12.2 Night backgrounds

The night minimum backgrounds in the 115 to 380 nm region are shown in Figure 12.1. In the nightime case, the minimum background in this region is at midlatitudes, between the auroral emission at higher latitudes and the tropical UV airglow found near the magnetic equator.

Hydrogen Lyman alpha emission at 121.6 nm dominates the night background throughout the ultraviolet region. It is typically about one-tenth of the day level, or between one and two kiloRayleighs. The other weak emissions in the MUV and NUV are from the nitric oxide delta and gamma bands and from molecular oxygen emission, primarily in the Herzberg bands. These molecular emissions are due to chemiluminescent reactions, and they have been discussed in detail in Chapter 10.

Night EUV backgrounds from the P78-1 satellite, *Chakrabarti et al.*, 1984, are less than 10 R and are seen for only the brightest features in Table 12.2. The O I recombination continuum extending to wavelengths shorter than 91.1 nm is the brightest feature, with an intensity of 35 R.

Although night backgrounds are much lower than day backgrounds, the limb enhancement effect should be kept in mind, since some emissions are fairly localized in altitude. The O_2 Herzberg bands when observed at the

182 CHAPTER 12. ATMOSPHERIC ULTRAVIOLET BACKGROUNDS

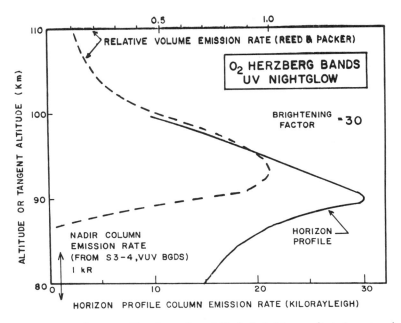

Figure 12.8: Oxygen Herzberg band limb brightness (total system)

limb maximum may be comparable to the brightest atmospheric radiance features, as displayed in Figure 12.8. Note however that these calculations are for the total emission system and that most spectral bandwidths would be smaller. In this figure, all of the observed oxygen band systems in the MUV and NUV are considered together and called the Herzberg system.

12.3 Auroral backgrounds

The ultraviolet experimenter should always keep in mind that the auroral region is a major contributor to the background. It is variable in strength, global location, and spectrum, but it has been found that the auroral oval precipitation region is always present someplace in the polar or even mid-latitude region, depending on solar activity.

The principal discussions of the aurora are in Chapter 10, which includes tables of spectra, and in Chapter 16. The aurora also appear slightly different in different wavelength bands, as is illustrated in Chapter 10 with Polar BEAR satellite simultaneous multispectral images. Most measurements of the UV auroral background from space are in the FUV (100 to 200 nm), and the comments here largely apply to that region.

While the auroral zone is variable, and will be different for every pass of a satellite, some general guidelines can be given. The nadir auroral intensity maximum in the auroral oval, as seen from space, is usually about the same as the local day airglow background. This behavior is seen in the full orbit records of Figures 12.4, 12.5, 12.6, and 12.7. The local day airglow background at the high latitudes of the typical aurora is greatly reduced from the maximum day nadir background found with an overhead sun, because of the larger solar zenith angle, as can readily be seen from the photometer full orbit scans, which include the low SZA case seen near the equator.

The day and night auroral displays have somewhat different characteristics, due to differences in the precipitating particles. The night auroral oval emission in the FUV dominates the region, as no other comparable source of background radiation is in the airglow. The oxygen resonance multiplet at 130.4 nm is a very sensitive indicator for the presence of the aurora, as it is undetectable in the nearby night airglow at typical auroral latitudes in the S3-4 data records.

For analysis of space images, the aurora is generally considered to come from an E-region altitude of about 110 km. However, this is a gross approximation, as illustrated dramatically by photographs of the auroral display from high inclination orbits of the shuttle. These photographs show the emission along striated columns of emission extending to altitudes much above the E-region. Color film shows a change from red to greenish-white from high to lower altitudes. There are no doubt similar differences in the ultraviolet emission. There are no known auroral UV limb scans that have been published to date, but the extensive observations taken on the STS-39 flight of the shuttle in April, 1991, may result in some.

12.4 Clutter

Clutter is the name given to nonstatistical local variability in a background measurement. As used here, clutter is the result of geophysics rather than statistics, and it generally determines the ultimate limit for detection. An attempt to measure clutter should be made to rule out this noise source for the spatial scale of interest in any given application. Statements that there are no known physical processes leading to small scale clutter should not be taken seriously, in the absence of measurements at the spatial scales of interest.

The detection of a target above a background level is ultimately dependent on the variability or clutter in the background. This variability determines the extent to which the background can be "zeroed out" so that the target can be detected in its presence. The concern is that the clutter may be the most important factor in determining the signal-to-noise ratio, (defined in Chapter 3).

There are no definitive measurements of clutter in the ultraviolet at small spatial scales (meters). Examinations of the Polar BEAR imager data, which has a limiting spatial resolution of about 10 km, reveals no clutter above the statistical level. Estimated UV clutter levels from the Polar BEAR auroral imager operated in the photometer mode (see Chapter 16), allow an analysis at the 10 to 100 meter resolution level, *Wohlers et al.*, 1989, but the counting statistics are a limitation, as shown in power spectral density plots. A discussion of the spatial characteristics of the background using Polar BEAR data is given by *Huguenin et al.*, 1989. The UV clutter problem is discussed also by *Malkmus et al.*, 1989.

Clutter has been found to be very important in *infrared* applications, and the interested reader is referred to the book by *Spiro and Schlessinger*, 1989, for a discussion of atmospheric infrared clutter. They divide the clutter problem into spatial, temporal, and jitter-induced clutter. Jitter is motion about the nominal line of sight due to vibration of the observing platform or movement within the deadband of a tracker. It is likely that UV clutter will need to be studied as thoroughly as IR clutter, as applications are developed requiring small instantaneous fields of view.

Clutter measurements seek to isolate the spatial and the temporal clutter components. Ideally, the two are separated by staring at a given field of view, or scene, over a period of time. Alternatively, clutter may be studied by making the measurements rapidly enough so that the field of view is not appreciably changed during the time it takes to record a series of frames of the scene from a moving platform such as a satellite.

The situation is complicated because low altitude satellites are moving rapidly relative to the Earth location. This motion is roughly 7.5 km/sec along the ground track. Studies of the requirements for fixed nadir satellite observations make it apparent that sensors with apertures much larger than the typical 20 to 40 cm diameter will be needed to collect enough counts for measurements of spatial scales in the meter class. Large apertures translate into large sensors, large costs, and limited flight opportunities. The inter-

12.4. CLUTTER

ested reader can make these comparisons using the backgrounds data and equations given in this book.

The staring approach is difficult also, because the airglow, scatter, and auroral radiance sources making up the background are volume or extended sources. Thus, they are generally distributed throughout broad altitude regions in the stratosphere, mesosphere, and thermosphere, rather than being due to the Earth's surface and cloud layers, as is usually the case in the IR. The typical IR approach of staring at a ground scene is not appropriate for the UV. A tangent altitude of interest can be identified, and this tangent altitude can be used to point a staring sensor, but the transmission path and solar conditions will change for the typical observation, making definitive analysis from a clutter point of view always somewhat incomplete.

While the complete scientific characterization of UV clutter may thus present difficulties, useful background clutter measurements associated with specific observing scenarios can be readily obtained. Thus, the observation of a moving discrete emission source (target) against the extended background radiance source from a moving platform can be done either by *tracking* the object, which involves keeping it fixed in the field of view of a sensor, or by *staring* at a field to watch the object move through it.

The tracking background can be called a *fast background*, because the changing FOV sweeps rapidly over different background scenes. The staring background can be called a *slow background*, since the field is more localized and would be expected to change less over the engagement time. Measurements of the fast and slow background clutter for important observing situations could well isolate important factors for detailed scientific analysis later, while at the same time providing the engineering data needed to assess the importance of this potential problem for high spatial resolution applications.

Clutter measurements must be made with spectral and spatial resolutions better than those used for most measurements made to date. For the spectral case, out-of-spectral band response from scattered visible sunlight or nearby brighter UV features must be eliminated. This "red leak," or long wavelength tail response may be removable with special solar blind filters or dual spectrometers, but with a great loss in sensitivity. The sensitivity can in principle be restored with a larger aperture, for the same angular resolution, but eventually practical constraints on space vehicles, as discussed in Chapter 5, will be exceeded. The spatial resolution needed may also require special baffles to give low response from directions beyond the design FOV.

Small comet absorption effects have been reported in the DE-1 FUV images of the Earth dayglow by *Frank et al.*, 1986a,b. From our point of view, these "holes" in the day background could be called a type of clutter in atmospheric remote sensing applications. It can be calculated that in the geologic time frame, these proposed comets would be the source of most of the water found today on the Earth, as discussed by *Frank*, 1990, in a popular account of this work.

However, the water comets are not frequent enough on our usual spatial and temporal scales to be a significant factor in our applications. Indeed, the low altitude satellites HILAT and Polar BEAR have not observed these holes in the data base to date, possibly because of the limited global coverage at any one time of these imagers, or because of the great differences in altitudes.

The assignment of the DE-1 observations to water comets is not generally accepted by the geophysics community at this time, and there is extensive debate in the literature (see *Dessler*, 1991, for a summary of the arguments against small comets). Even if the water comet explanation is eventually completely discredited and withdrawn, the source of the emission decreases in the DE-1 images needs to be found, as it will be of concern to experimenters and applications designers for future space measurements.

For most current UV applications in atmospheric remote sensing, which do not require extremely small spatial scales, clutter is not a factor. It does seem quite likely, however, that geophysical phenomena with small scale sizes will lead to clutter in ultraviolet backgrounds. These may well be discovered and studied as some applications requiring high spatial resolution lead to higher resolution imaging of the background radiance. Key geophysical suspects associated with clutter might be the gravity waves and wind shears that occur in the upper atmosphere, as these are known to affect the visible and infrared day airglow.

12.5 References

Anderson, Jr., D. E. and D. J. Strickland, Modeling UV-visible radiation observed from space, in *Atmospheric Propagation in the UV, Visible, IR, and MM-Wave Regions and Related Systems Aspects, Conf. Proc. 254*, AGARD, NATO, 35-1 to 35-11, 1989. (Available from NTIS, 5285 Port Royal Road, Springfield, VA 22161.)

Chakrabarti, S., R. Kimble, and S. Bowyer, Spectroscopy of the EUV (350–

12.5. REFERENCES

1400 Å) nightglow, *J. Geophys. Res., 89*, 5660–5664, 1984.

Chakrabarti, S., F. Paresce, S. Bowyer, R. Kimble, and S. Kumar, The extreme ultraviolet day airglow, *J. Geophys. Res., 88*, 4898–4904, 1983.

Dessler, A. J., The small comet hypothesis, *Rev. Geophysics, 29*, 355–382, 1991.

Fishburne, E. S., C. R. Waters, D. L. Moyer, and B. S. Green, OAO, A-2 measurements of the ultraviolet airglow, Technical Report to Air Force Cambridge Research Laboratory (now Phillips Laboratory), AFCRL-72-0318, 1972.

Frank, L. A., with P. Huyghe, *The Big Splash*, Carol Publishing Group, New York, 1990.

Frank, L. A., J. B. Sigwarth, and J. D. Cravens, On the influx of small comets into the Earth's upper atmosphere. I. Observations, *Geophys. Res. Lett., 13*, 303, 1986a.

Frank, L. A., J. B. Sigwarth, and J. D. Cravens, On the influx of small comets into the Earth's upper atmosphere. II. Interpretation, *Geophys. Res. Lett., 13*, 307, 1986b.

Henry, R. C., Ultraviolet background radiation, *Ann. Rev. Astron. Astrophys., 29*, 1991.

Huffman, R. E., Atmospheric emission and absorption of ultraviolet radiation, Chapter 22 in *Handbook of Geophysics and the Space Environment*, A Jursa, Scientific Editor, Air Force Geophysics Laboratory (now Phillips Laboratory), NTIS Document Accession Number ADA 167000, 1985.

Huffman, R. E., F. J. LeBlanc, J. C. Larrabee, and D. E. Paulsen, Satellite vacuum ultraviolet airglow and auroral observations, *J. Geophys. Res., 85*, 2201–2215, 1980.

Huffman, R. E., L. A. Hall, and F. J. LeBlanc, Comparison of the LOWTRAN 7 code and S3-4 satellite measurements, *Ultraviolet Technology III*, R. E. Huffman, Editor, *Proc. SPIE, 1158*, 38–44, 1989.

Huguenin, R., M. Wohlers, M. Weinberg, R. Huffman, R. Eastes, and F. DelGreco, Spatial characteristics of airglow and solar scatter radiance from the Earth's atmosphere, *Ultraviolet Technology III*, R. E. Huffman, Editor, *Proc. SPIE, 1158*, 16–27, 1989.

Malkmus, W., J. P. Filice, and C. B. Ludwig, Earthlimb radiance, transmissivities, and clutter, *Ultraviolet Technology III*, R. E. Huffman, Editor, *Proc. SPIE, 1158*, 69–83, 1989.

McPeters, R. D., Climatology of nitric oxide in the upper stratosphere, mesosphere, and thermosphere: 1978 through 1986, *J. Geophys. Res., 94*, 3461–3472, 1989.

Opal, C. B., Enhancements of the photoelectron excited dayglow during solar flares, *Space Research XIII*, Akademie-Verlag, Berlin, 1973.

Spiro, I. J. and M. Schlessinger, *Infrared Technology Fundamentals*, Dekker, 1989.

Strickland, D. J., R. P. Barnes, and D. E. Anderson, Jr., *UV Background Calculations: Rayleigh Scattered and Dayglow Backgrounds from 1200 to 3000 Å*, AFGL-TR-88-0200, Air Force Geophysics Laboratory, 1988.

Wohlers, M., R. Huguenin, M. Weinberg, R. Huffman, R. Eastes, and F. DelGreco, Estimated UV clutter levels at 10–100 meter sensor pixel resolution extrapolated from recent Polar BEAR measurements, *Ultraviolet Technology III*, R. E. Huffman, Editor, *Proc. SPIE, 1158*, 28–37, 1989.

Chapter 13

Radiance and Transmission Codes

Computational models and associated computer codes for ultraviolet radiance and transmission are under active development. These useful tools are now moving out of the hands of atmospheric researchers and are becoming available to a wider group of users. They enable a nonspecialist to obtain the radiance and transmission of UV in the atmosphere when the details of the problem, such as altitude, viewing direction, solar position, etc., are specified. The best models and codes encapsulate our current state of knowledge yet have the flexibility to easily accept new or different input parameters easily.

This chapter is an introduction to atmospheric radiance and transmission codes, beginning with the LOWTRAN series. It is possible to use LOWTRAN 7, the most recent version, for many UV problems. Development of a UV code called AURIC, which will include the FUV and EUV as well, is discussed. Some test cases demonstrating validation against real world data and a discussion of major uses of the codes as modules to be included in larger scene generators complete this chapter.

The scope here is limited to UV radiance and transmission codes. Computer models and codes have been applied to many remote sensing applications and instrumentation problems discussed in this book, as mentioned in appropriate chapters.

In order to be useful to a wide community of users, including nonspecialists, a code or model should be available to all qualified users through a central organization that exercises some control on the current best version of the code. The organization also interacts with users to produce improved versions as these become possible. The code should be nonproprietary, and the source code should be available. For some uses, such as in competitive evaluation of applications, it is necessary to have a standard code that can be specified for use in the preparation of proposals. These will doubtless be developed for the UV as they are now available in the IR and visible.

Commercial versions of atmospheric models and codes are available, which should increase their usage enormously. They are also appearing in versions that use personal computers and work stations. A government source of codes, software, and information about them is *Bilitza*, 1990.

13.1 LOWTRAN and similar user codes

The LOWTRAN series of computer models and codes has the ability to provide radiance and transmission in the longer wavelength regions of the ultraviolet, although certain limitations must be borne in mind. These codes were first developed to provide a simple and rapid means of obtaining accurate calculations of radiance and transmission over the range of 350 to 40,000 cm^{-1}, or to 250 nm in the UV, at the moderate resolution of 20 cm^{-1}. This resolution is about 0.1 nm in the MUV, which for many atmospheric problems would be considered reasonably high.

The origins of the LOWTRAN codes are described in the *Handbook of Geophysics and the Space Environment, Fenn et al.*, 1985. Also given in the handbook is the relationship of LOWTRAN to other codes such as FASCODE. The general idea of a low resolution transmission code in the UV, visible, and IR was given in a report by *Elterman*, 1968. LOWTRAN 2, described by *Selby and McClatchey*, 1972, provided an improved code for this purpose.

The series of codes has continued and been further improved by the addition of atmospheric radiance. The most recent versions are LOWTRAN 6, *Kneizys et al.*, 1983, and LOWTRAN 7, *Kneizys et al.*, 1988. Typically, each version retains the features of the previous ones, while adding new capabilities. These new capabilities result in part from the requests of users. LOWTRAN is widely accepted for use in infrared radiance and transmission problems (see, for example, *Spiro and Schlessinger*, 1989).

The extension of LOWTRAN 7 into the UV is described by *Anderson, et al.*, 1989. It now extends to 50,000 cm^{-1}, or 200 nm, while retaining the spectral resolution of 20 cm^{-1} (full width at half maximum) in steps of 5

cm^{-1}. Improved values for oxygen and ozone cross sections have been added. Other improvements are a revised solar spectrum and a better method for multiple scattering calculations.

Calculations of UV radiance from atmosphere scattering were discussed by *Green*, 1966, and *Dowling and Green*, 1966. These references also provide access to previous work. LOWTRAN 7 incorporates models of multiple scattering to a lower wavelength of 200 nm in a parameterization given by *Isaacs, et al.*, 1987.

The chief limitations for the UV are the restriction to an altitude of 100 km or below and the lack of airglow and fluorescence sources in the code. Nevertheless, if these factors are kept in mind, LOWTRAN can be very useful in obtaining estimates of UV radiance and transmission, *Dentamaro, et al.*, 1989.

The continuation of development of codes similar to LOWTRAN will be with MODTRAN, *Berk et al.*, 1989, which incorporates the capabilities of LOWTRAN but improves the resolution to 2 cm^{-1}, among other added features.

LOWTRAN 7 can be obtained on magnetic tape from the National Climatic Data Center, NOAA, Environmental Data Services, Federal Building, Ashville, NC 28801, (704) 259-0682. The licensed commercial PC version of LOWTRAN 7, called PCTRAN 7, is available from ONTAR Corporation, 129 University Road, Brookline, MA 02146-4532, (617) 277-6299.

Other UV codes are available. For modeling atmospheric propagation and lidar return in the visible and UV, the UVTRAN code has been described, *Patterson and Gillespie*, 1989. It is designed for relatively short ranges in the troposphere, and the aerosols parameterization approach used causes the major difference with LOWTRAN 7. The wavelength range is from 185 to 700 nm at a resolution of 2.5 nm for the absorber cross sections.

13.2 AURIC UV code

The AURIC code (Atmospheric Ultraviolet Radiance Integrated Code), now under development by the Geophysics Directorate, Phillips Laboratory, will be a radiance and transmission code for background calculations and remote sensing applications. The need to continue the development of radiance and transmission codes into the UV is apparent, as new remote sensing uses become of interest. This process is now beginning in the UV, as summarized in this section.

The elements of **phenomenology** needed for the code are largely available from the research results of the last few decades. Phenomenology is defined here as all of the scientific input needed for the models and codes. Where possible in AURIC, approaches taken for LOWTRAN and its successor MODTRAN will be used without change. For example, the LOWTRAN 7 atmospheric scattering modules are considered adequate for AURIC, as scattering is much less important at wavelengths less than 200 nm and at altitudes higher than 100 km.

For transmission, a central problem for atmospheric physics and chemistry in previous decades was the altitudes and effects of solar UV absorption in the upper atmosphere. As knowledge of the solar flux, photon cross sections, and atmospheric composition became better known, it was possible to make improved estimates of solar transmission and its effects, as shown by calculations of, for example, *Marr*, 1965, *Turco*, 1975, and *Simon and Brasseur*, 1983. LOWTRAN 7 has incorporated the transmission aspects of these phenomena to a lower wavelength of about 200 nm, and little or no modification is considered necessary. AURIC will continue this work into the FUV and EUV.

As mentioned above, LOWTRAN 7 does not include many UV radiance sources. These include airglow, molecular and atomic fluorescence scattering, the aurora, and possibly other sources. For these sources, what is sometimes called simply **chemistry** must be added to the models. In the most basic case, it involves a model that allows for all the reactions that occur among the constituents of the atmosphere. These include the excitation processes for airglows and auroral emissions and the emission rates themselves in the various radiative emission transitions. Examples of model calculations that address this part of the problem are given by *Anderson and Strickland*, 1988, 1989. Approaches similar to these will be included in AURIC.

The relationship between AURIC and the other items of technical work needed to develop, validate, and use the code are shown schematically in Figure 13.1.

The AURIC code is planned to use a resolution of 1 or 2 cm^{-1} and to concentrate initially on the 100 to 400 nm region. The wavelength region will ultimately extend into the visible and the EUV.

Some additional attributes of an ultraviolet radiance and transmission code such as AURIC should be as follows:

- The altitude range should extend from ground level to about 700 km, or to an altitude where there are no radiance sources or transmission

13.2. AURIC UV CODE

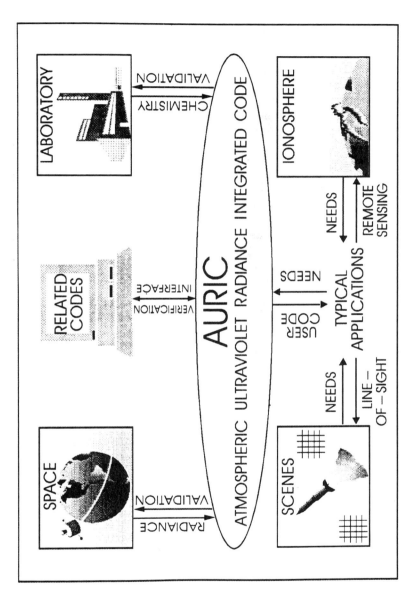

Figure 13.1: AURIC development, validation, and use

losses. The hydrogen Lyman α geocorona, (Chapter 11), is a special case involving higher altitudes.

- The solar UV input ought to be extended into the EUV to provide the excitation energy sources for some UV airglows; for example, the nitrogen Lyman-Birge-Hopfield bands in the FUV and MUV.

- To the maximum extent possible, atmospheric radiance sources should be calculated with *ab initio* methods rather than being included in look-up tables or as parameterizations.

- There should be provision for the addition of user-provided data such as revised model atmospheres and improved cross-section measurements.

- It should be a "living" code, supported by an organization which undertakes to maintain the standard version of the code, to correct mistakes, and to make improvements in it.

- To the maximum extent possible, it should be able to call up or be called up by other, similar codes in adjacent wavelength regions. The transitions between the codes should be as seamless as possible.

- It should initially be a line-of-sight code, with any position or altitude a possible location for either end of the path. All viewing directions including those toward the Earth, the Earth limb, and space should be possible.

- It should be readily incorporated into scene generators and test beds giving multidimensional simulations.

- It should be useful for any solar position relative to the line-of-sight, and for the complicated twilight situation, where the path may cross the terminator.

- It should cover all latitudes, with the simpler nonauroral case treated first and the auroral case covered eventually.

Considering the above desirable properties of a UV code and what is currently available in codes such as LOWTRAN 7 and MODTRAN, several specific technical areas are of high priority for early inclusion in AURIC. These technical areas include:

13.2. AURIC UV CODE

Oxygen Schumann-Runge bands The greatest challenge to extending codes to wavelengths shorter than 200 nm has been the accurate introduction of absorption in the Schumann-Runge bands, which have numerous narrow rotational lines that until recently had not been measured accurately. References to the photon cross sections have been given in Chapter 8. The problem has attracted many workers with several models and calculations, including *Hudson and Mahle*, 1972; *Park*, 1974; *Fang et al.*, 1974; *Kockarts*, 1976; *Strobel*, 1976; *Blake*, 1979; *Nicolet and Peetermans*, 1980; *Allison et al.*, 1986, (continuum); and *Nicolet and Kennes*, 1989. The approach favored for AURIC is to use fixed frequency increments rather than a line-by-line approach. It is necessary to include the temperature dependence of the cross section, which enormously complicates the problem.

Airglow excitation models The emission resulting from photoelectron excitation, chemiluminescence, and recombination must be included. Chapter 9 discusses these sources. *Meier*, 1991, has reviewed the current status of knowledge of airglow excitation. These radiance sources represent possibly the largest contributers not currently addressed in some fashion in LOWTRAN 7. The airglow models should cover solar zenith angle variation and include both day and night airglows.

Fluorescence For limb viewing directions and at twilight, sources such as the NO gamma and delta bands are important, as discussed in Chapter 11. A recent calculation provides some important cases involving this radiance source, *Strickland et al.*, 1988.

Resonance line multiple scattering The most intense radiance sources in the FUV are the H Lyman alpha line at 121.6 nm and the oxygen resonance multiplet at about 130.4 nm. They have been discussed in Chapter 11. The review by *Meier*, 1991, is an especially good introduction to the problems with calculation of these radiance sources. They should be included in codes, however, even if parameterizations have to be used, as their presence greatly influences sensor design, due to the scattered light that can be inadvertently introduced.

Atmospheric model The LOWTRAN code allows use of the US Standard Atmosphere, 1976; several atmospheres covering summer-winter and arctic-midlatitude-tropical cases; and user-provided model atmospheres. Where possible, the introduction of more detailed models such as MSIS 86 is desirable. Such models will be used, where the added complexity and possible effects on running time can be tolerated. Chapter 6 gives more information on these models.

Solar flux and variability As the major energy input into the atmosphere, accurate solar flux measurements must be used in the codes. *Anderson et al.*, 1989, discuss the solar spectrum used for LOWTRAN 7. Choices for AURIC are being evaluated. In addition, some account must be taken of the several kinds of solar variability that have been identified (see Chapter 7).

Twilight effects The difficult problems of lines of sight that cross or are near the terminator will be included in the code. *Anderson and Lloyd*, 1990, present detailed calculations at twilight illustrating the problems.

Auroral zone The emission from the aurora and polar cap structures described in Chapters 10 and 17, respectively, must eventually be included in models. Initial work will be concentrated on the nonauroral parts of the atmosphere. Some use of flags to alert the user to possible auroral effects may prove useful in initial models. Such flags could be tied to input values of K_p, which can provide a rough estimate of the location of the auroral oval.

The above lists of desirable properties will no doubt have to be modified in practice for a number of reasons. It is the best strategy to develop the code in such a way that it grows out of the current codes. In addition, there must be a rough parity among the various components of the code, so that one aspect of the code is not developed in great detail, while other parts are handled crudely.

The running time of the code must be considered. As computational power increases, more detailed codes will be possible, but the needs for rapid computation by users must be retained. All of these factors, as well as the needs mentioned at the beginning of this chapter should be considered in the development of improved UV codes.

13.3 Validation and verification

Validation is defined here as the comparison of a code calculation with measurements obtained in the "real world". **Verification** is similar, except

13.3. VALIDATION AND VERIFICATION

it is the comparison of one code with another to check that they give the same answer, within the approximations appropriate to the problem.

Validation can include both laboratory and field experiments. The validation of the code against accepted measurements provides confidence to users that it may be applied to other situations. It also can uncover cases where the codes and measurements are not in agreement. In these cases, either the code requires modification, more experiments are going to be required, or both code and experiment need improvement. Strictly speaking, a code is never fully validated, but the degree of validation can be judged by a potential user.

The measurements of UV backgrounds (airglow, aurora, scattering) and solar flux transmissions have been discussed in other chapters. Although much remains to be measured, much is already known. It is possible to begin the process of validation using LOWTRAN 7 by comparing field and laboratory measurements.

Anderson, et al., 1989, present comparisons of LOWTRAN 7 with balloon, rocket, and satellite data. Use is made of the code to a lower wavelength of about 192 nm using cross sections and solar fluxes that are in the code, although the code is fully functional only to 200 nm. A comparison with balloon solar flux data is shown in Figures 13.2, where the experimental data given by *Anderson and Hall*, 1986, is used. The solar irradiance at 33 km with a solar zenith angle of $24°$ is shown from calculation and measurement. The dominance of the oxygen Schumann-Runge bands from (1,0) to (4,0) is apparent. In Figure 13.2, the agreement is excellent, although close examination will reveal areas where improvements should be sought.

Also given by *Anderson et al.*, 1989, is a comparison of LOWTRAN 7 calculations with S3-4 satellite measurements of the day solar scatter background, *Huffman et al.*, 1980. This comparison is shown in Figures 13.3 and 13.4. The two panels show data from a spectrometer observing in the nadir direction from an altitude of about 200 km, which is well above the scattering layer. The minimum due to ozone absorption near 250 nm and the influence of solar Franhofer lines are apparent. The two comparisons are made at the low solar zenith angle (SZA) of $24°$ and the twilight SZA of $94°$.

As can be seen, the calculations give an excellent estimate of the radiance in Rayleigh/Å for the first case. At twilight, with the upper part of the atmosphere illuminated and the lower part in shadow, the gamma bands of nitric oxide due to solar fluorescence become the dominant emission features. These are not modeled by the code at this time and thus the code gives a poor estimate in this case. This difference between code and experiment indicates

CHAPTER 13. RADIANCE AND TRANSMISSION CODES

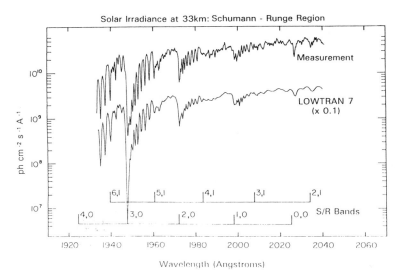

Figure 13.2: Comparison of balloon measurements and LOWTRAN 7 calculations of solar flux in region of oxygen Schumann-Runge bands

that fluorescence radiance must be added to the code. The comparison of LOWTRAN 7 with S3-4 measurements is discussed further by *Huffman et al.*, 1989.

The above examples illustrate uses of LOWTRAN 7 at stratospheric and higher altitudes. It is very desirable to have the code be valid at ground level, and in general the LOWTRAN 7 code is applicable to these problems. Continuing work on validation is needed, however. It has been pointed out by *Trakhovsky et al.*, 1989a, 1989b, that absorption by the Herzberg bands of molecular oxygen should be added to LOWTRAN. This very weak absorption occurs in the approximate range of 240 to 280 nm.

A final point should be made. Computer codes can become overloaded with radiance sources, reactions, and other details that have been included for "completeness". From time to time, it is necessary to conduct what are sometimes called *sensitivity studies*. A systematic effort is made to find and eliminate parts of the input that have little or no effect on the output of the code. While the sensitivity study and code revision has to be done carefully, some work of this nature is needed to keep a living code responsive to the needs of the users.

13.3. VALIDATION AND VERIFICATION

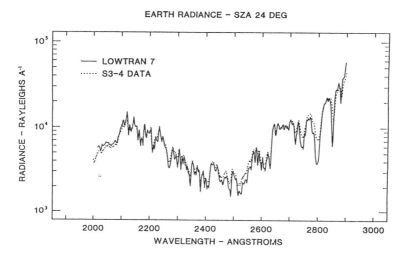

Figure 13.3: Comparison of LOWTRAN 7 calculation and S3-4 satellite measurements of the UV background, SZA 24° (near overhead sun)

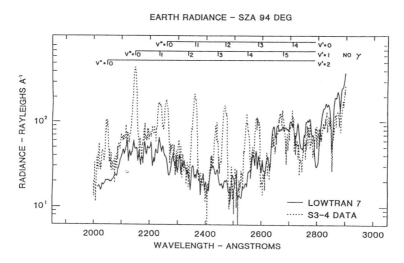

Figure 13.4: Comparison of LOWTRAN 7 calculation and S3-4 satellite measurements of the UV background, SZA 94° (twilight)

13.4 Related codes

The atmospheric radiance and transmission codes described in this chapter can be used in a stand-alone mode to provide answers to problems posed by users. They have other uses as well in relation to larger codes and models. For example, one use of these line-of-sight codes is as a module in a **scene generator**. The scene generator gives a two-dimensional image along a viewing direction, to simulate the FOV of a missile defense sensor, for example. The scene generator can be made realistic by inclusion of point sources of radiance and all the geophysical parameters known to be important. The uses of a scene generator range from system concept development to training of systems operators.

A more elaborate computer model, which may simulate theater or global operations, may be called a **test bed**. Scene generators and radiance and transmission codes may be used as some of the modules comprising a test bed. Just as with hardware, the overall capability of a large model depends on the quality of the component parts, which in this case includes the radiance and transmission codes.

While the emphasis in this book is on the ultraviolet, it must be kept in mind that there is no rigid barrier between the UV, visible, and IR. The physical processes do change gradually across the region, however. For example, much of the IR radiance is from atmospheres at local thermodynamic equilibrium, or LTE. The reader should be aware that UV radiance sources are non local thermodynamic equilibrium, or NLTE, sources.

Eventually, it will be useful to have radiance and transmission codes covering all of these wavelengths and accessible through one master code, or "shell," for convenience. For example, the MODTRAN and AURIC codes previously mentioned will eventually be combined with the codes SHARC, *Sharma et al.*, 1989, and AARC, *Winick et al.*, 1987. These codes cover limb and auroral radiances, respectively, in the IR and visible. The result could eventually be a general purpose radiance and transmission code from the EUV to the LWIR, or long wavelength infrared.

The current plan at the Phillips Laboratory, Geophysics Directorate, is to combine all the atmospheric radiance and transmission codes together with the celestial background codes under a shell called SMAART, or Smart Model for Atmospheric and Astronomical Radiance and Transmission. An effort will be made to have versions of this code that can run with desktop computers and work stations, in order to maximize the use of the codes.

A major use of UV radiance and transmission codes, or portions of them, will be for ionospheric electron density codes and atmospheric composition and density codes based on airglow, which are discussed in Chapters 14 through 17. Radiance and transmission codes can also be used in the global circulation models described briefly in Chapter 6.

13.5 References

Allison, A. C., S. L. Guberman, and A. Dalgarno, A model of the Schumann-Runge continuum of O_2, *J. Geophys. Res., 91*, 10,193–10,198, 1986.

Anderson, D. E. and S. A. Lloyd, Polar twilight UV-visible radiation field, Perturbations due to multiple scattering, ozone depletion, stratospheric clouds, and surface albedo, *J. Geophys. Res., 95*, 7429–7434, 1990.

Anderson, D. E. and D. J. Strickland, Synthetic dayglow spectra and the Rayleigh scattering background from the Far UV to the visible, *Ultraviolet Technology II*, R. E. Huffman, Editor, *Proc. SPIE 932*, 170–178, 1988.

Anderson, D. E. and D. J. Strickland, Modeling UV-visible radiation observed from space, *Conference Proceedings No. 254, Atmospheric Propagation in the UV, Visible, IR, and MM-Wave Region and Related Systems Aspects, AGARD, NATO* 35-1 to 35-11, 1989 (Available from the National Technical Information Service (NTIS), 5285 Port Royal Road, Springfield, VA 22161).

Anderson, G. P. and L. A. Hall, Stratospheric determination of O_2 cross-sections and photodissociation rate coefficients: 191–215 nm, *J. Geophys. Res., 91*, 14509–14514, 1986.

Anderson, G. P., F. X. Kneizys, E. P. Shettle, L. W. Abreu, J. H. Chetwynd, R. E. Huffman, and L. A. Hall, LOWTRAN 7 spectral simulations in the UV, *Conference Proceedings No. 254, Atmospheric Propagation in the UV, Visible, IR, and MM-Wave Region and Related Systems Aspects, AGARD, NATO*, 25-1 to 25-9, 1989 (Available from the National Technical Information Service (NTIS), 5285 Port Royal Road, Springfield, VA 22161).

Berk, A., L. S. Bernstein, and D. C. Robertson, *MODTRAN, A Moderate Resolution Model for LOWTRAN 7*, GL-TR-89-0122, Geophysics Laboratory (AFSC), 1989.

Bilitza, D., Solar-terrestrial models and application software, NSSDC WDC-A-R & S 90-19, National Space Science Data Center, NASA Goddard SFC, 1990.

Blake, A. J., An atmospheric absorption model for the Schumann-Runge bands of oxygen, *J. Geophysical Research, 84*, 3272–3282, 1979.

Dentamaro, A. V., C. G. Stergis, and V. C. Baisley, Minimizing the effects of the atmosphere in the observation of ultraviolet radiation, *Conference Proceedings No. 254, Atmospheric Propagation in the UV, Visible, IR, and MM-Wave Region and Related Systems Aspects, AGARD, NATO*, 5-1 to 5-11, 1989 (Available from the National Technical Information Service (NTIS), 5285 Port Royal Road, Springfield, VA 22161).

Dowling, J. and A. E. S. Green, Second-order scattering contributions to the Earth's radiance in the middle ultraviolet, Chapter 6 in *The Middle Ultraviolet: Its Science and Technology*, A. E. S. Green, Editor, Wiley, 1966.

Elterman, L., *UV, Visible, and IR Attenuation for Altitudes to 50 km*, AFCRL, Environmental Research Paper No. 285, 1968.

Fang, T.-M., S. C. Wofsy, and A. Dalgarno, Opacity distribution functions and absorption in Schumann-Runge bands of molecular oxygen, *Planet. Space Sci., 22*, 413–425, 1974.

Fenn, R. W., S. A. Clough, W. O. Gallery, R. E. Good, F. X. Kneizys, J. D. Mill, L. S. Rothman, E. P. Shettle, and F. E. Volz, Optical and infrared properties of the atmosphere, Chapter 18 in *Handbook of Geophysics and the Space Environment*, A. S. Jursa, Scientific Editor, Air Force Geophysics Laboratory, ADA 167000, 1985.

Green, A. E. S., The radiance of the Earth in the middle ultraviolet, Chapter 5 in *The Middle Ultraviolet: Its Science and Technology*, A. E. S. Green, Editor, Wiley, 1966.

Hudson, R. D. and S. H. Mahle, Photodissociation rates of molecular oxygen in the mesosphere and lower thermosphere, *J. Geophys. Res., 77*, 2902–2914, 1972.

13.5. REFERENCES

Huffman, R. E., F. J. LeBlanc, J. C. Larrabee, and D. E. Paulsen, Satellite vacuum ultraviolet airglow and auroral observations, *J. Geophys. Res., 85*, 2201–2215, 1980.

Huffman, R. E., L. A. Hall, and F. J. LeBlanc, Comparison of the LOWTRAN-7 code and S3-4 satellite measurements of UV radiance, *Ultraviolet Technology III*, R. E. Huffman, Editor, *Proc. SPIE 1158*, 38–44, 1989.

Isaacs, R. G., W.-C. Wang, R. S. Worsham, and S. Goldenberg, Multiple scattering LOWTRAN and FASCODE models, *Applied Opt., 26*, 1272–1281, 1987.

Kneizys, F. X., E. P. Shettle, W. O. Gallery, J. H. Chetwynd, L. W. Abreu, J. E. A. Selby, S. A. Clough, and R. W. Fenn, *Atmospheric Transmittance/Radiance: Computer Code LOWTRAN 6*, AFGL-TR-83-0187, Environmental Research Papers No. 846, 1983.

Kneizys, F. X., E. P. Shettle, L. W. Abreu, J. H. Chetwynd, G. P. Anderson, W. O. Gallery, J. E. A. Selby, and S. A. Clough, *Users Guide to LOWTRAN 7*, AFGL-TR-88-0177, Environmental Research Paper No. 1010, 1988.

Kockarts, G., Absorption and photodissociation in the Schumann-Runge bands of molecular oxygen in the terrestrial atmosphere, *Planet. Space Sci., 24*, 589–604, 1976.

Marr, G. V., The penetration of solar radiation into the atmosphere, *Proc. Roy. Soc., A, 288*, 531–539, 1965.

Meier, R. R., Ultraviolet spectroscopy and remote sensing of the upper atmosphere, *Space Sci. Rev., 58*, 1–185, 1991.

Nicolet, M. and R. Kennes, Aeronomic problems of molecular oxygen photodissociation VI. Photodissociation frequency and transmittance in the spectral range of the Schumann-Runge bands, *Planet. Space Sci., 37*, 459–492, 1989.

Nicolet, M. and W. Peetermans, Atmospheric absorption in the O_2 Schumann-Runge band spectral range and photodissociation rates in the stratosphere and mesosphere, *Planet. Space Sci., 28*, 85, 1980.

Park, J. H., The equivalent mean absorption cross sections for the O_2 Schumann-Runge bands: Application to the H_2O and NO photodissociation rates, *J. Atm. Sci.*, *31*, 1893–1897, 1974.

Patterson, E. M. and J. B. Gillespie, Simplified ultraviolet and visible wavelength atmospheric propagation model, *Applied Opt.*, *28*, 425–429, 1989.

Selby, J. E. A. and R. A. McClatchey, *Atmospheric Transmittance from 0.25 to 28.5 µm: Computer Code LOWTRAN 2*, AFCRL-TR-72-0745, AD 763 721, 1972.

Sharma, R. D., A. J. Ratkowski, R. L. Sundberg, J. W. Duff, L. S. Bernstein, P. K. Acharya, J. H. Gruninger, D. C. Robertson, and R. J. Healey, *Description of SHARC, the Strategic High-Altitude Radiance Code*, GL-TR-89-0229, Geophysics Laboratory (AFSC), 1989.

Simon, P. C. and G. Brasseur, Photodissociation effects of solar UV radiation, *Planet. Space Sci.*, *31*, 987–999, 1983.

Spiro, I. J. and M. Schlessinger, *Infrared Technology Fundamentals*, Dekker, 1989.

Strickland, D. J., R. P. Barnes, and D. E. Anderson, Jr., *UV Background Calculations: Rayleigh Scattered and Dayglow Backgrounds from 1200 to 3000 Å*, AFGL-TR-88-0200, Air Force Geophysics Laboratory, 1988.

Strobel, D. F., Parameterization of the atmospheric heating rate from 15 to 120 km due to O_2 and O_3 absorption of solar radiation, *NRL Memorandum Report 3398*, Naval Research Laboratory, Washington, D. C., 1976.

Trakhovshy, E., A. Ben-Shalom, and A. D. Devir, Measurements of tropospheric attenuation in the solar-blind UV spectral region and comparison with LOWTRAN 7 code, *Ultraviolet Technology III*, R. E. Huffman, Editor, *Proc. SPIE 1158*, 357–365, 1989a.

Trakhovshy, E., A. Ben-Shalom, U. P. Oppenheim, A. D. Devir, L. S. Balfour, and M. Engel, Contribution of oxygen to attenuation in the solar blind UV spectral region, *Applied Opt.*, *28*, 1588–1591, 1989b.

Turco, R. P., Photodissociation rates in the atmosphere below 100 km, *Geophysical Surveys*, *2*, 153–192, 1975.

Winick, J. P., R. H. Pickard, R. A. Joseph, R. D. Sharma, and P. P. Wintersteiner, *AARC, The Auroral Atmospheric Radiance Code*, AFGL-TR-87-0334, Air Force Geophysics Laboratory, 1987.

Chapter 14

Ozone and Lower Atmospheric Composition

Depletion of stratospheric ozone as a subject of concern to the health of the biosphere and its inhabitants needs no emphasis here. What may not be recognized is that the methods used to measure the global change of ozone composition are largely UV methods. Indeed, stratospheric ozone measurements are clearly the most significant application of UV techniques to atmospheric remote sensing to date.

This chapter describes UV methods for sensing constituents of the lower atmosphere, including the *troposphere*, extending from the ground to about 18 kilometers altitude, and the *stratosphere*, extending from about 18 to about 50 kilometers altitude (see Chapter 6). While the major emphasis is on **stratospheric ozone**, methods for minor and pollutant species are also included (Section 14.4).

The regions of the ultraviolet primarily involved include the near ultraviolet (NUV), from about 400 to about 300 nanometers, and the middle ultraviolet (MUV), from about 300 to about 200 nanometers.

This chapter draws on Chapters 7, 8, and 11, and the references therein. Additional general references to tropospheric chemistry and photochemistry are by *Finlayson-Pitts and Pitts*, 1986; *Graedel*, 1985; and *Warneck*, 1988. The *Handbook of Geophysics and the Space Environment* has an introduction to atmospheric composition by *Anderson et al.*, 1985, which includes many of the minor and pollutant species. The photochemistry of the stratosphere has been discussed by *Turco*, 1985.

Volcano dust, and also gaseous emissions such as SO_2, can also be tracked and measured by satellite sensors, including some of the ozone sensors now flying. Volcanic emissions pose a threat to aircraft flight and can greatly perturb the lower atmosphere.

In order to be given serious consideration in regard to long term global changes, the composition measurements being made at this time must be to a much higher degree of accuracy and precision than was typical in earlier research. Intercomparisons of methods and calibrations become very important. This requirement presents a challenge to our best technology, as the measurements must receive acceptance by governments and ultimately by people. We scentists have to learn to explain, to place our findings in context, and to have our work scrutinized. As war is too important to be left to generals, so now global change science may be too important to be left exclusively in the hands of scientists.

14.1 Atmospheric ozone

In the last twenty years, the study of ozone has changed from an academic specialty to a subject influencing great projects such as supersonic transports, *Johnston*, 1971, and leading to controls on the release of seemingly innocuous convenience chemicals such as chlorofluorocarbons, *Molina and Rowland*, 1974. The discovery of the ozone "hole" in the antarctic, *Farman et al.*, 1985, and later in the arctic has led to extensive scientific expeditions involving ground, airborne, and space measurements (see, for example, *Solomon*, 1988).

The importance of atmospheric ozone in determining the ultraviolet limit to solar radiation at ground level has been known for more than a century, as summarized by *Mitra*, 1947. Briefly, *Hartley*, 1881, suggested ozone as the cause in reporting the absorption bands in the 210 to 320 nm region that now bear his name. The argument became more convincing after *Huggins*, 1890, found additional absorption bands at wavelengths longer than 320 nm. Conclusive demonstration of the importance of ozone in establishing the ultraviolet limit was found by *Fowler and Strutt*, 1917. As Huggins had done previously, they used the difference in ultraviolet absorption spectra observed using as the source the star Sirius seen through varying atmospheric paths. Methods similar to this are still in use today.

A chapter by *Griggs*, 1966, describes the more recent understanding of atmospheric ozone, especially as it relates to the ultraviolet. The book by

14.1. ATMOSPHERIC OZONE

Brasseur and Solomon, 1984, and other books on the chemistry and physics of the troposphere and stratosphere cover continuing progress in the field. The amount of effort on the ozone "problem" is massive, and no attempt to cover it completely will be made here. Both ozone destruction and the effects on man are covered by *Bower and Ward*, 1982. Examples of recent conference proceedings will demonstrate the interest both for the study of ozone, *Bojkov and Fabian*, 1989, and for the health and environmental implications of ozone depletion, *Russell Jones and Wigley*, 1988.

Especially recommended as a reference to all methods of atmospheric ozone measurement is the December, 1989, special issue of the journal *Planetary and Space Science*, called the G. M. B. Dobson: Atmospheric Ozone Issue, *Farman and London*, 1989.

Instrumentation is covered in the very useful book by *Grant*, 1989, on ozone measurements in the stratosphere and troposphere. The importance of precision and accuracy is stressed, and reprints of the most significant papers are included.

Tropospheric ozone is a trace constituent of air near ground level due to the interaction of manmade or natural pollutants and sunlight in regions where the air circulation is poor. The buildup of photochemical "smog" is recognized as an important human health and agricultural problem. Ozone is a toxic gas, causing respiratory problems and eye irratation. Ozone levels in the troposphere are elevated when photochemical smog levels are significant. Ozone concentrations approaching several tenths of a part per million trigger governments to institute emission control regulations and issue public health advisories. The chemistry of photochemical smog and associated experimental methods is covered in detail in the text by *Finlayson-Pitts and Pitts*, 1986.

Ground level ozone can be measured using the attenuation of an ultraviolet light source due to ozone absorption over a fixed path length. A frequent choice is the use of the strong mercury resonance line at 253.7 nm for the light source. This emission line, readily available in mercury arc light sources, is near the peak of the strong Hartley absorption band of ozone. Measurements over pathlengths of up to several kilometers are possible at and near ground level. Further details of this commonly used method are given in reviews by *Ayers and Gillett*, 1990, and by *Grant*, 1989. Sensors are described by *Platt and Perner*, 1983, and by *Proffitt and McLaughlin*, 1983.

The reader should remember that the tropospheric **bad ozone** and stratospheric **good ozone** are separate problem areas, with different sets of formation and destruction processes. Bad ozone contaminates the air we breathe, while good ozone protects us from damaging solar UV radiation. Our principal concern is with the good ozone in the stratosphere.

Stratospheric ozone makes up about 90% of the ozone in any vertical column of the atmosphere. Ozone formation in the stratosphere results from atomic oxygen reactions through the well-known Chapman mechanism, *Chapman*, 1930. Atomic oxygen at these altitudes is formed by photoabsorption:

$$O_2 + h\nu\,(\lambda < 242nm) \longrightarrow O + O.$$

There are three main reaction paths for the atomic oxygen. It may recombine with another atomic oxygen in the presence of a third body M:

$$O + O + M \longrightarrow O_2 + M.$$

The atomic oxygen may react with molecular oxygen, also in the presence of a third body, to form ozone:

$$O + O_2 + M \longrightarrow O_3 + M.$$

Finally, the atomic oxygen may react with and destroy an ozone molecule:

$$O + O_3 \longrightarrow O_2 + O_2.$$

Photodissociation is also an important natural destruction reaction (see Chapter 8). In the absence of significant destruction of ozone by minor species, the steady state levels of ozone are determined by these reactions.

Further discussion of the many catalytic cycles also occurring in the stratosphere is beyond our scope. This is currently a very active area of research, with reactive Cl and Br species of several types implicated in ozone destruction. These species, which may be more efficient than atomic oxygen in destroying ozone, are increasing due to atmospheric pollution. The concern is that there will be an increase of harmful UV radiation reaching the surface.

The many chemical reactions and photochemical processes associated with stratospheric ozone that cannot be reviewed here are covered in the references. The emergence of standard sources of rate constants and photon cross sections is a valuable product of work on the ozone problem. The periodically updated reviews of *DeMore et al.*, 1990, are recommended.

14.1. ATMOSPHERIC OZONE

One new avenue in ozone research is the recognition of the importance of heterogeneous reactions in the stratosphere. These reactions occur at the surface of aerosols, but they are difficult to study in the laboratory compared to gas phase homogenous reactions. Another new aspect is the recognition that there are a whole range of sources of gases from agricultural, forestry, and industrial activities that have been neglected in earlier work. These must all be considered and incorporated into models.

Stratospheric ozone has a complicated global formation and circulation pattern. It is produced most abundantly in equatorial regions, where the photodissociation of O_2 is strongest. The ozone then tends to circulate to the polar regions, where larger column densities occur. The global zonal average total ozone concentrations on a monthly basis are shown in Figure 14.1. An introduction to the global dynamics of ozone distribution is given by *Salby and Garcia*, 1990, who stress the importance of planetary waves in transporting, or advecting, the ozone from the tropics to the poles.

The total column density of ozone is frequently given in **Dobson units (D.U.)**, which are the number of milliatmosphere centimeters of ozone at standard temperature of 0 °C and standard pressure of 1 atmosphere (abbreviated STP). A midrange column density is 300 D.U., which would correspond to a total column density of ozone equivalent to 0.3 cm of ozone at STP. The total D.U. range usually encountered is from about 200 to 600 D.U., with larger values occurring in the polar winter. However, the reported minimum in the antarctic ozone hole in 1991 was about 110 D.U. In terms of molecules, using Loschmidt's number of $2.687 \times 10^{19} \, molec./cm^3$ at STP, 300 D.U. is equivalent to $8.06 \times 10^{18} \, molec./(cm^2)(column)$. The peak concentration in the stratosphere, between 20 and 30 km, is about $6 \times 10^{12} \, molec./cm^3$. Other ways of specifying ozone concentrations include the use of mixing ratios at specific atmospheric pressures and densities.

All ultraviolet methods for atmospheric ozone measurement are based at least in part on absorption. The bands and absorption cross sections were introduced in Chapter 7. Briefly, the strong Hartley band absorption in the 200 to 310 nm region and the weaker Huggins bands in the 310 to 350 nm region are utilized. The weak Chappius bands between 450 and 750 nm in the visible region and infrared emission near 9.6 micrometers are also used in remote sensing techniques for ozone. Other ozone measurements from space use microwave techniques.

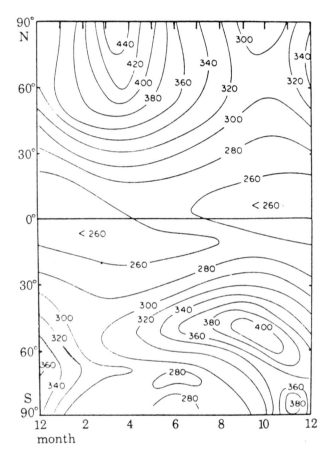

Figure 14.1: Global zonal total ozone concentrations (Dobson units)

14.2 Local stratospheric ozone measurements

Many stratospheric ozone techniques give measurements at or near a fixed geographic location. These include methods using ground, aircraft, balloon, or sounding rocket platforms. The techniques will be described in the approximate order that they were introduced. A network of local observing stations can of course approximate worldwide coverage.

PASSIVE ABSORPTION

These methods depend on directly observing the absorption of sunlight or another source. A review of these methods, which were among the first to be used to study atmospheric ozone, has been given by *Grant*, 1989.

Ground-based differential absorption The Dobson spectrophotometer in various forms has been used since the 1920s to obtain ozone column densities, *Komhyr et al.*, 1989. The M-83 filter ozonometer, *Gushchin et al.*, 1985, and the Brewer spectrophotometer, *Kerr et al.*, 1985, are considerably different in detail but are based on similar principles. The instruments observe either scattered solar radiation or direct solar or lunar absorption. Wavelengths or pairs of wavelengths in the 290 to 340 nm region are used, with special corrections for unwanted scattering by molecules and aerosols in the atmosphere. The variation with altitude is possible using the *umkehr* technique. A recent assessment of errors and accuracies in zenith measurements has been given *Stamnes et al.*, 1990. The use of UV stars of known spectral brightness as sources for absorption in the Huggins bands can be used.

Direct in-situ absorption The availability of high altitude balloons and sounding rockets has led to the direct use of absorption to obtain not only the column density but also improved altitudinal distributions. Straightforward application of the Beer-Lambert absorption law discussed in Chapter 5 is possible. The chaotic nature of the detailed distribution due to local transport processes such as winds and planetary waves has become apparent in these measurements. Use of sounding rockets for this purpose has been reported from the early 1950s, as recently reviewed by *Smith and Hunten*, 1990. A current version of the technique is the ROCOZ-A, *Barnes and Simeth*, 1986. Most of these measurements use the sun or a star as the source. An interesting use of the night airglow as the light source to measure mesospheric ozone has been reported by *Reed*, 1968. Also in the category of in-situ absorption is a balloon-borne instrument that carries its own light source for absorption measurements of ozone, *Proffitt and McLaughlin*, 1983.

LIDAR METHODS

Methods to probe the stratosphere and troposphere are being established using laser backscatter, or LIDAR, methods. Because of their power and weight requirements, lidar methods are best suited to ground and aircraft use. The differential absorption lidar, or DIAL, technique uses pairs of nearby wavelengths beamed in the zenith direction from lasers. One wavelength is strongly absorbed and the other is weakly absorbed, but they are affected similarly by scattering so that a correction can be made in the return signals. A suitable pair of wavelengths obtainable from a xenon chloride laser is 308 nm for the absorbing wavelength and 353 nm for the reference wavelength. For an example of the use of the method see *Megie et al.*, 1985. Two detailed reports on recent installations are given by *McDermid et al.*, 1991, and *McGee et al.*, 1991.

14.3 Global stratospheric ozone measurements

Detailed global study and monitoring of ozone is done now largely by sensors carried on Earth orbiting satellites. At this time, over two decades of measurements are available. Limitations have been the cost and complexity of satellites, the lack of hands-on calibration and maintenance of equipment, and the normal gestation period needed for the adoption of any new technology. These problems have all been generally overcome, and satellite measurements are vital to current ozone research.

Reviews of satellite methods for ozone measurement have been given by *Krueger et al.*, 1980, and by *Miller*, 1989, who also cover infrared methods. Satellite ozone measurements involving the ultraviolet are discussed further here as either *occultation* or as *solar backscatter* methods.

OCCULTATION METHODS

Occultation is the loss of intensity from an observed star or the sun due to atmospheric absorption while it is being eclipsed by the Earth or another planet. The analysis of these intensity records from a satellite enables an altitude profile of an absorbing species to be obtained. The exponential

14.3. GLOBAL STRATOSPHERIC OZONE MEASUREMENTS

decrease in density with altitude results in most absorption occurring at the *tangent altitude*, or the altitude of closest approach of the line of sight to the solid Earth (see Chapter 5). A recent review of occultation techniques and their use in studying the atmospheres of the Earth and other planets was given by *Smith and Hunten*, 1990, who also discuss refractive effects and other sources of uncertainty in the method.

Application of the stellar occultation approach to measurement of ozone from a satellite has been by *Riegler et al.*, 1976, 1977, from the OAO-3 satellite. Extensive use of the method has been on the Stratospheric Aerosol and Gas Experiment (SAGE I and II). This sensor uses four bands near 385, 450, 600, and 1000 nm to obtain ozone, nitrogen dioxide, and aerosol altitude distributions. SAGE I operated from February 1979 to December 1984, and SAGE II has been in operation since October 1984. The Chappius band absorption of ozone in the visible wavelength region has also been used, *Chu and McCormick*, 1979.

The Ultraviolet Spectrometer and Polarimeter (UVSP) on the Solar Maximum Mission uses occultation at 279 nm in the Hartley continuum for mesospheric ozone measurements, *Aikin et al.*, 1982. It began operation in November, 1984. Results from UVSP and SAGE II are in reasonable agreement in altitude regions where they overlap, *Aikin et al.*, 1989. The comparison of measurements from SAGE II and from ROCOZ-A sounding rockets has recently been reported, *Barnes et al.*, 1991.

The proposed GOMOS instrument is planned to continue measurement of ozone with stellar occultation using wavelengths in the 250 to 675 nm range, *Bertaux et al.*, 1991.

SOLAR BACKSCATTER METHODS

The determination of the daytime ozone concentration in the atmosphere by measurement of backscattered solar radiation has become possibly the most widely used technique to measure and monitor ozone levels from satellites. The scattered solar radiation levels in the Hartley and Huggins band regions (roughly 200 to 350 nm) are decreased by absorption due to ozone. This leads to the minimum in the atmospheric background centered near 250 nm seen from space, as shown in Figure 12.1. From the radiance measurements, both the **ozone profile**, or number density as a function of altitude, and **total ozone**, or the number of molecules in a $cm^2 column$, can be measured.

The use of backscatter methods was suggested several times when satellite experimentation became feasible, *Singer*, 1956; *Singer and Wentworth*,

1957; *Twomey*, 1961; *Green*, 1964, 1966; and *Krasnopol'skiy*, 1966. Backscatter measurements appear to have been done first in the UV by *Rawcliffe and Elliot*, 1966, using a single band at 284 nm in the Hartley band absorption region. There have been many satellite experiments and dedicated satellite programs since this time, and the reader will find detailed reviews in *Krueger et al.*, 1980, and in *Miller*, 1989.

Multipurpose research satellites that have reported ozone measurements include OGO-4, *Anderson et al.*, 1969; Solar Mesospheric Explorer, *Barth et al.*, 1983; *Rusch et al.*, 1983, 1984; and Dynamics Explorer, *Keating et al.*, 1985. The UV instrumentation used by these programs are discussed elsewhere in this book (see Section 5.5).

Dedicated, or operational, programs intended to provide long-term measurements have been in operation for a number of years. These programs are usually referred to by the names of their sensors, which include the Backscattered Ultraviolet (BUV) experiment, the Solar Backscattered Ultraviolet Radiometer (SBUV), and the Total Ozone Mapping Spectrometer (TOMS). There are also more recent improved versions, such as the SSUV/2 and a version for the shuttle called the SSBUV.

The satellite ozone measurement programs discussed in this section have the common thread of being on sun-synchronous *Nimbus* or *NOAA* satellites that are in circular orbits at altitudes of between about 950 and 1100 km. The polar orbit generally has a midday sun synchronous local time. Total ozone is obtained by all of these sensors, and all but TOMS obtains the ozone profile between about 25 and 55 km. In addition to the ultraviolet radiance, the solar flux is measured at each wavelength with the same photomultiplier detector by use of a diffuser plate.

The BUV and SBUV sensors have been described by *Heath et al.*, 1973, 1975. The SBUV obtains measurements at 12 wavelengths between 250 and 340 nm using a bandpass of 1 nm. The fixed wavelengths are scanned in a step-scan mode. The wavelengths from 250 to 306 nm are used for ozone profiles, and wavelengths between 312 and 340 nm are used for total ozone. The instrument consists of two 250 mm Ebert-Fastie spectrometers operated in tandem to reduce the out-of-band scattered light. The detector is an EMR 510 photomultiplier with a bialkali photocathode. Additional details are described in the references. The sensor obtains measurements along the satellite track in the nadir direction.

The TOMS sensor obtains images of the total ozone content of the atmo-

14.3. GLOBAL STRATOSPHERIC OZONE MEASUREMENTS

sphere. This sort of data has become of great interest, because total ozone maps of the globe can be obtained from a series of satellite orbits with periods of about 100 minutes. In contrast to the rapid (up to seconds) variability of the aurora, which can be imaged at FUV wavelengths (see Chapter 16), the ozone variability is slower. Complete global maps of the daytime total ozone can be obtained daily. TOMS sensors are or will be flown on a number of satellites, including the Soviet Union's Meteor-3 satellite. A recent review of TOMS data is by *Krueger*, 1989.

The TOMS sensor is described by *Heath et al.*, 1975. It is a simpler instrument than the SBUV, using a single 250 mm Ebert-Fastie spectrometer that approximates one-half of the SBUV. The spectrometer isolates six fixed wavelengths in the 313 to 380 nm range and measures the radiance with a photomultiplier similar to that used for the SBUV. The image is obtained by a cross track scanning mirror moving the observed area over a total swath of about $102°$ in $3°$ steps. This cross track is scanned in about eight seconds, and the field of view is about 50 km by 50 km at the nadir. A special filter is used to alleviate the effects of clouds, since at these wavelengths the observed radiance may originate from near or at the surface.

The data sets from both the SBUV and the TOMS approaches have received much attention. The error sources have been carefully analyzed, and most attention is now on the diffuser plate used for the solar flux measurement. This component changes in efficiency during the years of operation in flight, but work is progressing to alleviate this problem. All phases of the calibration of the sensor receive considerable attention, as is necessary in careful work of this type.

Ozone profile measurements give the number density of ozone as a function of altitude. These measurements are possible with the backscatter method because of the difference in scattering efficiency and photoabsorption. The observed radiance at these wavelengths in a 1 nm band comes from a series of fairly small altitude region of about 14 km in altitude, depending on the wavelength. Thus, the altitude region from 25 km to 55 km is scanned by using a series of wavelengths between about 350 and 250 nm, respectively, *Heath et al.*, 1973. The mathematical inversion techniques and algorithms used in the current programs can be found through the references.

Total ozone measurements give the total column amount of ozone above ground level. The method uses the longer wavelength range of approximately 310 to 380 nm, which penetrates to ground level. It was pointed out by *Dave and Mateer*, 1967, and later demonstrated with BUV data by *Mateer et al.*,

1971, that such a measurement was possible. The approach, using pairs of wavelengths, has similarity to the Dobson approach used in ground level local ozone measurements. It is also readily combined with imaging, as discussed earlier for the TOMS sensor.

Extensive data bases of satellite observations now exist from about 1970. These data bases may be accessed through organizations such as the National Space Science Data Center, Goddard SFC, Greenbelt, Maryland.

Concern for atmospheric changes over time has led to extensive analysis of all aspects of these measurements, as discussed in many places. A recent example, among many, is by *Chandra et al.*, 1990.

The development of satellite measurement of ozone from initial space experimentation to its present state is a useful example of how **technology transition** occurs in practice. The refinement of well-understood techniques, the adoption of excellent calibration methods, and the reliance on proven sensor detectors and spectrometers are evident. This development provides a model for ultraviolet auroral imaging and global weather systems development, which are currently at the experimental stage (see Chapters 16 and 17).

14.4 Minor species

Ultraviolet methods are used to find the concentrations of several important minor constituents of the troposphere and stratosphere. Experimental methods, including UV methods, used in the troposphere have been reviewed by *Ayers and Gillett*, 1990. Instrumentation for both tropospheric and stratospheric measurement of chemical species has recently been reviewed by *Kolb*, 1991.

Ultraviolet techniques have been adopted to measure several mi nor species. Depending on their chemical and photochemical reactivities, these can frequently be much more important than their number density might suggest. For example, they may be catalysts in the destruction of ozone.

Recent references to passive ultraviolet methods from the ground or aircraft cover measurement of hydroxyl radical, OH, *Burnett and Burnett*, 1983, 1984, *Burnett et al.*, 1990, *Wang et al.*, 1981, *Dorn et al.*, 1988; nitrogen dioxide, NO_2, *Wahner et al.*, 1990; chlorine dioxide, OClO, *Schiller et al.*, 1990; bromine monoxide, BrO, *Wahner et al.*, 1990; and chlorine monoxide, ClO, *Burnett*, 1989.

Thus, some of the key intermediates in ozone destruction by chlorofluorocarbons can be measured with these ultraviolet methods. Techniques to discriminate against the effects of aerosols and of Rayleigh scattering may be needed, but the methods acquire valuable measurements to aid in understanding levels of ozone and other atmospheric constituents.

In addition to the above methods, there are in-situ measurement techniques for stratospheric minor species that use UV in a number of ways. The most recent description of these measurements and their significance in regard to ozone destruction is due to *Anderson et al.*, 1991. In this case, an aircraft in the stratosphere samples the ambient composition of free radicals containing Cl and Br through use of fast chemical reactions and UV resonance line scattering by the atoms. Other UV related techniques have been used by this group to sample the atmosphere flowing through a reaction chamber after it is dropped from a balloon.

Finally, volcanic particulate and SO_2 emission can be measured and tracked using the TOMS ozone sensor from satellites, *Doiron et al.*, 1991.

14.5 References

Aikin, A. C., B. Woodgate, and H. J. P. Smith, Atmospheric ozone determination by solar occultation using the UV spectrometer on the Solar Maximum Mission, *Appl. Opt., 21*, 2421, 1982.

Aikin, A. C., D. J. Kendig, and H. J. P. Smith, An intercomparison of mesospheric ozone profiles determined by the UVSP and SAGE II solar occultation experiments, *Planet. Spa. Sci., 37*, 97–104, 1989.

Anderson, G. P., C. A. Barth, F. Cayla, and J. London, Satellite observations of the vertical ozone distribution in the upper stratosphere, *Ann. Geophys., 25*, 341, 1969.

Anderson, G. P., H. S. Muench, R. E. Good, C. R. Philbrick, and W. Swider, Atmospheric Composition, Chapter 21 in *Handbook of Geophysics and the Space Environment*, edited by A. S. Jursa, pp. 21-1 to 21-67, Air Force Geophysics Laboratory, NTIS ADA 167000, 1985.

Anderson, J. G., D. W. Toohey, and W. H. Brune, Free radicals within the antarctic vortex: The role of CFCs in antarctic oxone loss, *Science, 251*, 39–46, 1991.

Ayers, G. P. and R. W. Gillett, Tropospheric chemical composition: Overview of experimental methods in measurement, *Rev. Geophysics, 28*, 297–314, 1990.

Barnes, R. A. and P. G. Simeth, Design of a rocket-borne radiometer for stratospheric ozone measurements, *Rev. Sci. Instrum., 57*, 544, 1986.

Barnes, R. A., L. R. McMaster, W. P. Chu, M. P. McCormick, and M. E. Gelman, Stratospheric Aerosol and Gas Experiment II and ROCOZ A ozone profiles at Natal, Brazil: A basis for comparison with other satellite instruments, *J. Geophys. Res., 96*, 7515–7530, 1991.

Barth, C. A., D. W. Rusch, R. J. Thomas, G. H. Mount, G. J. Rottman, G. E. Thomas, R. W. Sanders, and G. M. Lawrence, Solar Mesosphere Explorer: Scientific objectives and results, *Geophys. Res. Lett., 10*, 237-240, 1983.

Bertaux, J. L., G. Mégie, T. Widemann, E. Chasseféire, R. Pellinen, E. Kyrola, S. Korpela, and P. Simon, Monitoring of ozone trend by stellar occultations: The GOMOS instrument, *Adv. Space Res., 11*, 327–342, 1991.

Bojkov, R. D. and P. Fabian, Editors, *Ozone in the Atmosphere, Proceedings of the Quadrennial Ozone Symposium 1988 and the Tropospheric Ozone Workshop*, Deepak Publishing, Hampton, Va., 1989.

Bower, F. A. and R. B. Ward, editors, *Stratospheric Ozone and Man*, Volumes I and II, CRC Press, 1982.

Brasseur, G. and S. Solomon, *Aeronomy of the Middle Atmosphere*, Reidel, 1984.

Burnett, C. R. and E. B. Burnett, OH Pepsios, *Appl. Optics, 22*, 2887–2892, 1983.

Burnett, C. R. and E. B. Burnett, Observational results on the vertical column abundance of atmospheric hydroxyl: Description of its seasonal behavior 1977–1982 and of the 1982 El Chichon perturbation, *J. Geophys. Res., 89*, 9603–9611, 1984.

Burnett, C. R., K. R. Minschwaner, and E. B. Burnett, OH vertical column abundance: Tropical measurements, *J. Geophys. Res., 95*, 16,491–16,495, 1990 and references therein.

14.5. REFERENCES

Burnett, E. B., Fourier-analytic technique for the separation of the signature of atmospheric ClO absorption from the solar background spectrum in the near ultraviolet, *Appl. Optics, 28*, 430–436, 1989.

Chandra, S., R. D. McPeters, R. D. Hudson, and W. Planet, Ozone measurements from the NOAA-9 and the Nimbus-7 satellites: Implications of short and long term variabilities, *Geophys. Res. Lett., 17*, 1573–1576, 1990.

Chapman, S., On ozone and atomic oxygen in the upper atmosphere, *Phil. Mag., 10*, 369, 1930.

Chu, W. P. and M. P. McCormick, Inversion of stratospheric aerosol and gaseous constituents from spacecraft solar extinction data in the 0.38–1.0 micron wavelength region, *Appl. Opt., 18*, 1404, 1979.

Dave, J. V. and C. L. Mateer, A preliminary study of the possibility of estimating total atmospheric ozone from satellite measurements, *J. Atmos. Sci., 24*, 414–427, 1967.

DeMore, W. B., S. P. Sander, D. M. Golden, M. J. Molina, R. F. Hampson, M. J. Kurylo, C. J. Howard, and A. R. Ravishankara, *Chemical Kinetics and Photochemical Data for Use in Stratospheric Modeling, Evaluation Number 9*, JPL Publication 90-1, Jet Propulsion Laboratory, Pasadena, California, 1990.

Doiron, S. D., G. J. S. Bluth, C. C. Schnetzler, A. J. Krueger, and L. S. Walter, Transport of Cerro Hudson SO_2 clouds, *EOS, Trans, Am. Geophys. Soc., 72*, 489, 1991.

Dorn, H. P., J. Callies, U. Platt, and D. H. Ehhalt, Measurements of tropospheric OH concentrations by laser long-path absorption spectroscopy, *Tellus, 40B*, 437–445, 1988.

Farman, J. C. and J. London, Guest Editors, *G. M. B. Dobson: Atmospheric Ozone Issue, Planet. Space Sci., 37*, 1485- 1672, 1989.

Farman, J. C., B. G. Gardiner, and J. D. Shanklin, Large losses of total ozone in Antarctica reveal seasonal ClO_x/NO_x interaction, *Nature, 315*, 207, 1985.

Finlayson-Pitts, B. J. and J. N. Pitts, Jr., *Atmospheric Chemistry: Fundamentals and Experimental Technique*, Wiley, 1986.

Fowler, A. and Hon. R. J. Strutt, *Proc. Roy. Soc. A, 93*, 577, 1917.

Graedel, T. E., The photochemistry of the troposphere, Chapter 2 in *The Photochemistry of Atmospheres: Earth, the Other Planets, and Comets*, J. S. Levine, editor, 39–76, Academic Press, 1985.

Grant, W. B., editor, *Ozone Measuring Instruments for the Stratosphere*, Vol. 1 of Collected Works in Optics, Optical Society of America, Washington, D. C., 1989.

Green, A. E. S., Attenuation by ozone and the earth's albedo in the middle ultraviolet, *Appl. Optics, 3*, 203–208, 1964.

Green, A. E. S., *The Middle Ultraviolet: Its Science and Technology*, A. E. S. Green, editor, p. 127, Wiley, 1966.

Griggs, M., Atmospheric Ozone, Chapter 4 in *The Middle Ultraviolet: Its Science and Technology*, A. E. S. Green, editor, 83–117, Wiley, 1966.

Gushchin, G. P., S. A. Sokolenko, and V. A. Kovalyov, Total ozone measuring instruments used at the USSR station network, p. 543 in *Atmospheric Ozone*, C. S. Zerefos and A. Ghazi, editors, Reidel, 1985 (reprinted in *Grant*, 1989).

Hartley, W. N., *Jour. Chem. Soc., 39*, 57 and 111, 1881.

Hays, P. B., R. G. Roble, and A. N. Shah, Terrestrial atmospheric composition from stellar occultations, *Science, 176*, 793–794, 1972.

Heath, D. F., C. L. Mateer, and A. J. Krueger, The Nimbus 4 backscattered ultraviolet (BUV) atmospheric ozone experiment—two years operation, *Pure Appl. Geophys., 106–108*, 1238, 1973.

Heath, D. F., A. J. Krueger, H. A. Roeder, and B. D. Henderson, The solar backscatter ultraviolet and total ozone mapping spectrometer (SBUV/TOMS) for Nimbus G, *Opt. Eng., 14*, 323, 1975.

Huggins, M. L., *Proc. Roy. Soc. A, 48*, 216, 1890.

Johnston, H., Reduction of stratospheric ozone by nitrogen oxide catalysts from supersonic transport exhaust, *Science, 173*, 517, 1971.

14.5. REFERENCES

Keating, G. M., J. D. Craven, L. A. Frank, D. F. Young, and P. K. Bhartia, Initial results from the DE-1 ozone imaging instrumentation, *Geophys. Res. Lett.*, *12*, 593, 1985.

Kerr, J. B., C. T. McElroy, D. I. Wardle, R. A. Olafson, and W. F. J. Evans, The automated Brewer spectrophotometer, p. 396 in *Atmospheric Ozone*, C. S. Zerefos and A. Ghazi, editors, Reidel, 1985 (reprinted in *Grant*, 1989).

Kolb, C. E., Instrumentation for chemical species measurements in the troposphere and stratosphere, *Rev. Geophysics, Supplement; U.S. National Report to IUGG, 1987-1990*, 25–36, 1991.

Komhyr, W. D., R. D. Grass, and R. K. Leonard, Dobson spectrophotometer 83: A standard for total ozone measurements, 1962–1987, *J. Geophys. Res.*, *94*, 9847, 1989.

Krasnopol'skiy, V. A., Ultraviolet spectrum of the radiation reflected by the earth's atmosphere and its use in determining the total content and vertical distribution of atmospheric ozone, *Geomag. and Aeronomy*, *6*, 236–242, 1966.

Krueger, A. J., The global distribution of total ozone: TOMS satellite measurements, *Planet. Space Sci.*, *37*, 1555–1565, 1989.

Krueger, A. J., B. Guenther, A. J. Fleig, D. F. Heath, E. Hilsenrath, R. McPeters, and C. Prabhakara, Satellite ozone measurements, *Phil. Trans. Roy. Soc. London, A 296*, 191–204, 1980.

Mateer, C. L., D. F. Heath, and A. J. Krueger, Estimation of total ozone from satellite measurements of backscattered ultraviolet earth radiance, *J. Atmos. Sci.*, *28*, 1307–1311, 1971.

McDermid, I. S., D. A. Haner, M. M. Kleiman, T. D. Walsh, and M. L. White, Differential absorption lidar systems for tropospheric and stratospheric ozone measurements, *Opt. Engineering*, *30*, 22–30, 1991.

McGee, T. J., D. Whiteman, R. Ferrare, J. J. Butler, and J. F. Burris, STROZ LITE: Stratospheric ozone lidar trailer experiment, *Opt. Engineering*, *30*, 31–39, 1991.

Megie, G., G. Ancellet, and J. Pelon, Lidar measurements of ozone vertical profiles, *Appl. Optics*, *24*, 3454, 1985.

Miller, A. J., A review of satellite observations of atmospheric ozone, *Planet. Spa. Sci., 37*, 1539–1554, 1989.

Mitra, S. K., *The Upper Atmosphere*, The Royal Asiatic Society of Bengal, 1 Park Street, Calcutta, 16, see Chapter 4, The Ozonosphere, 1947.

Molina, M. J. and F. S. Rowland, Stratospheric sink for chlorofluorocarbons—chlorine atom catalyzed destruction of ozone, *Nature, 249*, 810, 1974.

Platt, U. and D. Perner, Measurements of atmospheric trace gases by long path differential long path UV/visible absorption spectroscopy, in *Optical and Laser Remote Sensing*, D. K. Killinger and A. Mooradian, editors, pp. 97–105, Springer-Verlag, New York, 1983.

Proffitt, M. H. and R. J. McLaughlin, Fast response dual-beam UV absorption ozone photometer suitable for use on stratospheric balloons, *Rev. Sci. Instruments, 54*, 1719–1728, 1983.

Rawcliffe, R. D. and D. D. Elliot, Latitude distribution of ozone at high altitudes, deduced from a satellite measurement of the earth's radiance at 2840 Å , *J. Geophys. Res., 71*, 5077–5089, 1966.

Reed, E. I., A night measurement of mesospheric ozone by observations of ultraviolet airglow, J. Geophys. Res., 73, 2951–2957, 1968.

Riegler, G. R., J. F. Drake, S. C. Liu, and R. J. Cicerone, Stellar occultation measurements of atmospheric ozone and chlorine from OAO-3, *J. Geophys. Res., 81*, 4997, 1976.

Riegler, G. R., S. K. Atreya, T. M. Donahue, S. C. Liu, B. Wasser, and J. F. Drake, UV stellar occultation measurements of nighttime equatorial ozone, *Geophys. Res. Letters, 4*, 145, 1977.

Rusch, D. W., G. H. Mount, C. A. Barth, G. J. Rottman, R. J. Thomas, G. E. Thomas, R. W. Sanders, G. M. Lawrence, and R. S. Eckman, Ozone densities in the lower mesosphere measured by a limb scanning ultraviolet spectrometer, *Geophys. Res. Lett., 10*, 241, 1983.

Rusch, D. W., G. H. Mount, C. A. Barth, R. J. Thomas, and M. T. Callan, Solar mesosphere explorer spectrometer: Measurements of ozone in the 1.0 to 0.1 mbar region, *J. Geophys. Res., 89*, 11677, 1984.

14.5. REFERENCES

Russell Jones, R. and T. Wigley, Editors, *Ozone Depletion: Health and Environmental Consequences*, Wiley, 1988.

Salby, M. L. and R. R. Garcia, Dynamical perturbations to the ozone layer, *Phys. Today*, 38–46, March, 1990.

Schiller, C., A. Wahner, U. Platt, H.-P. Dorn, and J. Callies, Near UV atmospheric absorption measurements of column abundances during Airborne Arctic Stratospheric Expedition, January-February 1989, II. OClO observations, *Geophys. Res. Lett.*, *17*, 501–504, 1990.

Singer, S. F., Geophysical research with artificial earth satellites, in *Advances in Geophysics*, *3*, H. E. Landsberg, editor, 301–367, Academic Press, 1956.

Singer, S. F. and R. C. Wentworth, A method for the determination of the vertical ozone distribution from a satellite, *J. Geophys. Res.*, *62*, 299–308, 1957.

Smith, G. R. and D. M. Hunten, Study of planetary atmospheres by absorptive occultations, *Rev. Geophysics*, *28*, 117–143, 1990.

Solomon, S., The mystery of the Antarctic ozone hole, *Rev. Geophys.*, *26*, 131, 1988.

Stamnes, K., S. Pegau, and J. Fredrick, Uncertainties in total ozone amounts inferred from zenith sky observations: Implications for ozone trend analyses, *J. Geophys. Res.*, *95*, 16,523–16,528, 1990.

Tisone, G. C., Measurements of the absorption of solar radiation in the region of 2150 Å by O_2 and O_3, *J. Geophys. Res.*, *77*, 2971, 1972.

Turco, R. P., The photochemistry of the stratosphere, Chapter 3 in *The Photochemistry of Atmospheres: Earth, the Other Planets, and Comets*, J. S. Levine, editor, 77–128, Academic Press, 1985.

Twomey, S., On the deduction of the vertical distribution of ozone by ultraviolet spectral measurements from a satellite, *J. Geophys. Res.*, *66*, 2153–2162, 1961.

Wahner, A., J. Callies, H.-P. Dorn, U. Platt, and C. Schiller, Near UV atmospheric absorption measurements of column abundances during Airborn Arctic Stratospheric Expedition, January-February 1989, I.

Technique and NO_2 observations, III. BrO observations, *Geophys. Res. Lett.*, *17*, I. 497–500, III. 517–520, 1990.

Wang, C.C., L. I. Davis, Jr., P. M. Selzer, and R. Munoz, Improved airborne measurements of OH in the atmosphere using the technique of laser induced fluorescence, *J. Geophys. Res.*, *86*, 1181-1186, 1981.

Warneck, P., *Chemistry of the Natural Atmosphere*, Academic Press, 1988.

Chapter 15

Upper Atmospheric Composition and Density

This chapter covers the principal ultraviolet methods used to find the number densities of the major constituents of the upper atmosphere, which includes the mesosphere and the thermosphere. The mesosphere, between about 50 and 85 kilometers, and the thermosphere, between about 85 and 1000 kilometers, have as their primary neutral constituents O_2, N_2, O, NO, N, H, and He (see Figure 6.2).

In the mesosphere and thermosphere, the absorption of solar radiation, the transmission from and through the region, and the remote sensing methods all tend to be in the far ultraviolet (FUV), from about 200 to 100 nm, and the extreme ultraviolet (EUV), from about 100 to 10 nm.

These UV remote sensing methods are primarily based on the *absorption, fluorescence, and airglow* properties of the constituents, which are sections of this chapter. These topics are introduced in Chapters 7, 8, 9, and 11. General references to this chapter are *Rees*, 1989, *Torr*, 1985, and *Grossman et al.*, 1987. An especially valuable recent paper is by *Meier*, 1991.

This chapter covers only the nonauroral atmosphere, but it must be kept in mind that both the aurora and the ionosphere are at mesospheric and thermospheric altitudes. The following two chapters cover applications to auroral imaging (Chapter 16) and ionospheric electron density sensing (Chapter 17).

Density for satellite orbit prediction is a major applications need, and ultraviolet remote sensing may well have a place in future measurements approaches, either through composition measurements or more indirectly through solar flux monitoring for global models. Section 15.5 discusses density and orbit prediction.

15.1 Absorption methods

Molecular oxygen (O_2), molecular nitrogen (N_2), and atomic oxygen (O), the primary gases in the mesosphere and lower thermosphere, can all be remotely sensed with ultraviolet absorption methods. All atmospheric constituents absorb at some ultraviolet wavelength, and the straightforward use of this property can be one of the most unambiguous methods to obtain atmospheric number density. The absorption equations in Chapter 7 and the absorption cross sections of Chapter 8 provide the basis for methods of this type.

A light source, such as the sun or a UV star, and a detector in the right position are all that are required to obtain the absorption data. The column density can be converted to local number density using the variation of intensity as a rocket rises through the atmosphere or as the light from an astronomical object is seen at different tangent altitudes from an orbiting platform. An obvious complication is that several or all constituents may absorb at some wavelengths, so that data from several wavelengths may be needed to separate the absorption from the different species.

The eclipse of the light from the sun or a star, usually called *occultation*, was discussed for ozone in the last chapter. During the occultation, the absorption is localized near the tangent altitude, or the altitude of closest approach to the Earth of the optical path from the source to the observer. Intense solar lines or UV stars can be used. The excellent review of occultation methods by *Smith and Hunten*, 1990, should be consulted for early measurements and an introduction to the analysis of occultation data, not only for the Earth but also for the other planets.

The use of stellar sources rather than the sun enables the altitude resolution to be improved. Stellar sources also allow the measurements to be made at any local time. In the solar case, the measurements are confined to near sunrise and sunset, where species concentrations are undergoing their maximum diurnal changes. It should also be noted that the precise determination of the tangent altitude is not a simple problem, and this determination may be the largest uncertainty in the measurement. Stellar occultation methods for remote sensing and various associated effects such as atmospheric refraction have been discussed by *Hays and Roble*, 1968a,b.

15.1. ABSORPTION METHODS

MOLECULAR OXYGEN: O_2

Atmospheric absorption using the sun as the source was studied in some of the first sounding rocket experiments. The discovery of significant molecular oxygen densities at altitudes up to 140 km was reported by *Friedman, et al.*, 1951. Simple photochemical models that excluded transport processes had predicted no molecular oxygen above about 100 km, and so the models required adjustment. The experiment was based on absorption in the oxygen Schumann-Runge continuum in the FUV.

Absorption spectra from the sun obtained by *Jursa et al.*, 1959, 1963, 1965 on sounding rocket flights were used to measure molecular oxygen and nitric oxide concentrations. While valid oxygen measurements were obtained, the nitric oxide absorption predicted by then current models was not found and only an upper limit was obtained. This was also an important observation, since atmospheric models had to be adjusted to accommodate unexpectedly low nitric oxide measurement. Further nitric oxide measurements have been made by the more sensitive fluorescence technique, as discussed in the next section. In these early experiments, spectrographs using photographic detection methods were used, which requires recovery of the payload and careful photometric techniques. Although photoelectric methods and telemetry are now used, the measurements themselves continue to be valid.

Molecular oxygen absorption in the Schumann-Runge continu a and bands has continued to be used widely for composition studies. For example, the far UV emission from the sun has been used by *Weeks and Smith*, 1968, *Norton and Warnock*, 1968, and *Weeks*, 1975a,b. Stellar occultation has been used by *Opal and Moos*, 1969, *Hays and Roble*, 1972, *Hays et al.*, 1972, and *Atreya et al.*, 1976. For the measurements in the Schumann-Runge continuum, it has been sufficient to use a detector covering a broad wavelength band such as a simple photometer or an ionization cell.

Molecular oxygen densities in the mesosphere have been measured with the use of solar hydrogen Lyman α absorption by *Prinz and Brueckner*, 1977. Absorption of solar radiation in the 215 nm region by both O_2 and O_3 has been used as the basis of a composition measurement by *Tisone*, 1972. The solar wavelengths of 133.5, 121.6, and 103.2 nm were used from the SKYLAB manned satellite to measure molecular oxygen concentrations in the thermosphere, *Garriott et al.*, 1977.

As emphasized by *Smith and Hunten*, 1990, most of these occultation measurements have not utilized instrumentation specifically designed for atmospheric composition. This comment applies both to the sensor itself and to the associated pointing and tracking methods of the space vehicle. Thus, considerable improvement in the technique could be expected using upgrades of the instrumentation.

The use of *day airglow* emission in the nitrogen Lyman-Birge-Hopfield (LBH) bands as a source for the remote sensing of O_2 by *absorption* has recently been considered. The intensities of the LBH bands observed either in the nadir or limb direction from space are modified by molecular oxygen absorption in the Schumann-Runge continuum. Use of this effect as a part of a remote sensing approach has been pointed out by *Meier and Anderson*, 1983, and possibly by others. Further discussion of LBH is in Section 15.3 on airglow methods.

EUV ABSORPTION: O_2, N_2, and O

The use of solar EUV for composition measurements was demonstrated in a series of sounding rocket experiments by *Hinteregger*, 1962, and *Hall et al.*, 1963, 1965. The latter paper describes use of a spectrometer measuring nine prominent solar emission lines between 120.6 and 30.4 nm. The absorption curves of these lines together with the absorption cross sections of molecular nitrogen, molecular oxygen, and atomic oxygen provides enough information to extract number densities for the separate constituents. An example of composition measurements made in this way is shown in Figure 15.1.

Satellite measurements were used to produce similar results, *Hinteregger and Hall*, 1969, using occultation data from an experiment primarily dedicated to measuring the solar flux and carried on the OSO-III satellite. Also from a similar satellite occultation experiment, the relative absorption cross sections for atomic oxygen in the EUV have been obtained, *Knight et al.*, 1972.

It would appear that the use of EUV occultation has not yet received the attention that it deserves, and it may prove useful for remote sensing when more solar EUV observations are conducted. Eventually, the use of stars as light sources in the EUV will enable this method to escape the criticism that the measurements are confined to near terminator conditions at local sunrise and sunset.

A difficulty with the absorption method, especially with the use of solar lines as the source, may be the need for accurate and detailed knowledge of emission and absorption line shapes. When an incident solar emission line is absorbed by the same transition that led to its emission, the line shapes

15.1. ABSORPTION METHODS

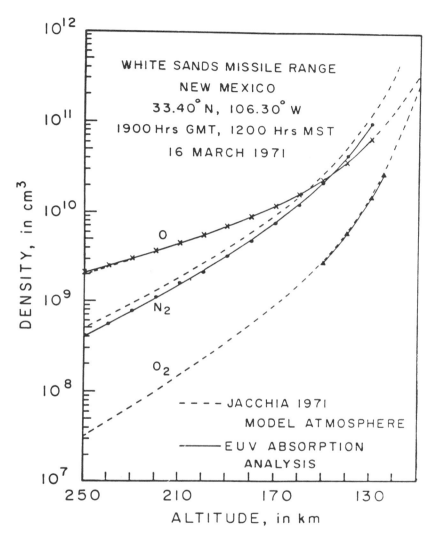

Figure 15.1: Thermospheric densities with EUV absorption (L. A. Hall)

of both emission and absorption transitions have to be considered. The situation is much more complicated than the simple exponential extinction law expressions of Eqs. 7.1 and 7.2. This approach was used in sounding rocket experiments that observed the profile of the broad solar H Lyman α line (121.6 nm) for sensing of **atomic hydrogen**, *Jones et al.*, 1970, and of **atomic deuterium**, *Bruner and Wilson*, 1969.

Knowledge of the spectroscopy of all atmospheric species is vital in avoiding difficulties in the use of absorption methods, particularly in the EUV. Atomic oxygen and atomic nitrogen have strong absorption lines, and molecular nitrogen has strong bands with narrow rotational lines in this region which may dominate other absorption when they overlap solar lines.

This possible complication with solar line sources was pointed out by *Huffman and Larrabee*, 1968, for several intense solar lines, including some considered by *Hall et al.*, 1965 for neutral composition measurements. A similar complication occurs in the use of the O^+ emission at 83.4 nm for ionospheric electron density profile measurements, as pointed out by *Cleary et al.*, 1989. Note that this complication may well be avoided through the use of stellar sources, which have additional choices for absorption wavelengths.

The great advantage of absorption and occultation methods from space is that the absolute sensitivity of the detector does not have to be known. Only a relative measurement of intensities is needed. Thus, concerns about calibration change in orbit, contamination of optics, and recalibration during flight are eliminated or greatly reduced. It is apparent that absorption and occultation methods should be considered carefully where they are a possibility.

15.2 Fluorescence and scattering methods

Nitric oxide (NO), atomic hydrogen (H), and atomic helium (He) can be measured remotely in the mesosphere and thermosphere by their ultraviolet fluorescence. *Atomic oxygen, O,* composition from fluorescence scattering has been extensively studied, but it is made so difficult by multiple scattering that it is not clear this radiation can be used for remotely sensing composition. Fluorescence and scattering in the atmosphere have been introduced in Chapter 11.

NITRIC OXIDE, NO, was suggested as a key minor constituent in the D region of the ionosphere by *Nicolet*, 1945. Its importance derives from the fact that it can be photoionized by the solar hydrogen Lyman alpha line,

15.2. FLUORESCENCE AND SCATTERING METHODS

which penetrates to about 70 km due to the fortuitous window in the molecular oxygen absorption cross section (see Chapter 8). A sounding rocket experiment to attempt to measure nitric oxide by absorption spectroscopy, *Jursa, et al.*, 1959, found instead an upper limit, indicating the need for more sensitive methods.

Barth, 1964, 1966, discovered and developed the use of the fluorescence of nitric oxide in the gamma bands in the region of 215 nm as a method of obtaining number densities in the mesosphere. This is now the standard method for nitric oxide densities in the daytime and twilight. Sounding rockets were used for the initial measurements of Barth and for many further investigation (see *Cleary*, 1986, for references).

It was found from nadir observing spectrometers that the gamma band emission could be observed under twilight conditions from the satellites OGO-4, *Rusch*, 1973, and S3-4, *Huffman et al.*, 1980.

Measurements at all local times can be found with limb observations from satellites, and this is the preferred approach for global sensing. Measurements have been reported from the Atmospheric Explorer (AE) satellite series by *Cravens et al.*, 1985, using methods originally described by *Cravens*, 1977.

An extensive set of NO measurements using limb fluorescence in the gamma bands has been obtained from the Solar Mesosphere Explorer (SME) satellite program, *Barth et al.*, 1983, and *Rusch et al.*, 1984. Applications of SME measurements to auroral storm effects and to solar cycle variability, respectively, have been reported by *Siskind et al.*, 1989a,b and by *Barth et al.*, 1988. A complete review of SME and AE nitric oxide measurements is beyond our scope, but these are available through the more recent references. Reference models for thermospheric NO are now available through satellite measurements, *Barth*, 1990. The SME ozone measurements are described in Chapter 14, and the discovery of polar mesospheric clouds by this satellite is discussed in Chapter 11.

ATOMIC HYDROGEN, H, present in the thermosphere and exosphere scatters solar radiation from the intense solar Lyman alpha emission line at 121.6 nm and from other members of this series. The name *geocorona* has been given to this fringe of hydrogen atoms surrounding the Earth. Geocoronal scattering of solar radiation is a major part of the background radiation of the Earth as seen from space. Nadir viewing measurements have found an intensity range of from 10 to 20 kiloRayleighs at low solar zenith angles in the daytime. At night, the intensity is still 1 to 2 kiloRayleighs, due to scattering from the dayside. The geocorona extends out for several

Earth radii into the exosphere.

Recent measurements and interpretations of this emission have been from the Dynamics Explorer I satellite by *Rairden et al.*, 1986, and from the 78-1 satellite by *Anderson et al.*, 1987. Another recent discussion is by *Gladstone*, 1988. The previous extensive study and modeling of this emission can be found through these references.

Lyman alpha scattering offers a method to obtain the atomic hydrogen number density in the thermosphere and exosphere. The experimental technology is available, and the models are in principal suitable for this application. More accessible user models however are needed. The emission varies with the solar cycle, and it could possibly find an application in monitoring solar flux variability from some spacecraft. Comprehensive atmospheric composition measurement programs should consider the use of the Lyman alpha emission for hydrogen atom number densities.

A secondary consideration for remote sensing in the far ultraviolet must always be to avoid unwanted interference from both hydrogen Lyman alpha and beta (121.6 and 102.5 nm). Previous observations and radiance models can sometimes be used to demonstrate in the sensor design that this radiation will not influence the desired measurement. However, the availability of user friendly radiance models for the hydrogen resonance lines would be of great value to provide more assurance that unwanted scattered light will not influence the measurements. Eventually, radiance sources of this type will be incorporated into the AURIC code (see Chapter 13).

ATOMIC HELIUM, He, in the thermosphere and exosphere scatters solar radiation at the 58.4 and 53.7 nm resonance lines. The stronger 58.4 nm line can be used for composition measurements, but use of the 53.7 nm line is complicated by nearby O II lines. Limb measurements are particularly valuable. There have been a number of measurements, with the major satellite observations and models reported by *Meier and Weller*, 1972, and *Anderson et al.*, 1979 from the STP 72-1 satellite and *Chakrabarti et al.*, 1983, from the STP 78-1 satellite. The use of an absorption cell to acheive high spectral resolution has recently been reported by *Fahr and Smid*, 1986.

ATOMIC OXYGEN, O, has a number of emission features in the FUV and EUV that have been considered for remote sensing. Apart from observational difficulties with interfering emission lines in crowded regions of the spectrum, the chief difficulty with their use appears to be the complications in the intensity and altitude profile of the observed emission due to strong multiple scattering of the emission before it reaches the detector. The situation is further complicated by the fact that the emission features may

15.2. FLUORESCENCE AND SCATTERING METHODS

be produced by both scattering of solar lines and photoelectron excitation.

The atomic oxygen emission from the multiplets near 130.4 and 135.6 nm are very prominent features of the far ultraviolet background. Indeed, for many observational situations, the 130.4 nm feature is larger than hydrogen Lyman alpha. It therefore may be the strongest backgrounds emission feature in the FUV and EUV, usually being greater than 10 kiloRayleighs. The 135.6 nm emission is usually several hundred Rayleighs.

The **OI 130.4 nm** emission is produced by both solar scattering and photoelectron excitation, with the relative amounts varying with geophysical conditions. Further, the atmosphere is extremely optically thick to the emitted radiation, so multiple scattering models have to be used. Detailed discussion is to be found in *Meier*, 1991, and the conclusion is that at this time the measurement of this radiation alone is not a promising remote sensing method for atomic oxygen number density. However, it should be pointed out that if modeling of the scattering could be understood better, there is probably much remote sensing information given by this multiplet and its line shape. Thus, use of the 130.4 nm emission for composition measurements possibly awaits improved data at higher spectral resolution and realistic models under the conditions of the measurements.

The **OI 135.6 nm** emission is very useful for remote sensing, as the day production mechanism is primarily from photoelectron excitation rather than from solar scattering. Since scattering is a minor concern for this line, it will be discussed further in the airglow methods section. Indeed, as will be seen, the use of the 135.6 nm airglow emission is key to many proposed atmospheric and ionospheric remote sensing strategies.

The weak **OI 164.1 nm** emission observed by the STP S3-4 satellite, *Huffman et al.*, 1980, or the ratio of this line to the 130.4 emission line, has been proposed as a remote sensing method, *Meier and Conway*, 1985, and *Conway et al.*, 1988. The ratio should be linear with the atomic oxygen concentration at least under some conditions. Both emissions are from the same upper level, but the 130.4 nm emission is several orders of magnitude brighter, making it necessary to have sensors capable of handling a large dynamic range. Uncertainty about how to account for the 130.4 scattering remains a problem with this approach.

The ratio **OI 117.3/OI 98.9 nm** has been proposed as a method following the observation of the weak 117.3 nm emission on a sounding rocket experiment, *Bowers et al.*, 1987. The 98.9 nm emission is a relatively strong resonance line. This ratio has been modeled by *Gladstone et al.*, 1987.

In summary, *Meier*, 1991, concludes that the 117.3/98.9 nm ratio is probably the best approach involving ratios of atomic oxygen emission lines after also considering 164.1/130.4 and 135.6/130.4. He also feels that all of these methods have difficulties usually associated with multiple scattering that may prevent their adoption as remote sensing methods. More measurements are indicated. A discussion of difficulties with use of the 117.3 and 164.1 nm lines and a set of useful laboratory spectra are given by *Erdman and Zipf*, 1987; however, the difficulties do not seem to be insurmountable. Further discussion of the UV dayglow from atomic oxygen for planetary structure and composition has been given by *Gladstone*, 1988.

In-situ atomic oxygen UV fluorescence has been used from sounding rockets as a sensing technique for atomic oxygen. While the measurement is local to the sounding rocket, the method depends on fluorescence from the same atomic oxygen excited states as those considered in this section. An on-board source of the 130.4 nm oxygen atom resonance multiplet illuminates the local atmosphere, and a detector also on the rocket measures the scattered signal. Complications with this method are due to the line shape from the light source and the Doppler shift of the line due to the velocity of the sounding rocket. *Sharp*, 1991, has recently reviewed and described this method in detail.

15.3 Airglow methods

Molecular nitrogen, N_2, and *atomic oxygen*, O, emission from the day airglow can be used to remotely sense these constituents. *Atomic oxygen*, O, and *atomic nitrogen*, N, can be determined from the nightglow of O_2 and NO. In contrast to the emission sources in the last section, which were due to solar scattering, airglow emission is the result of collisional or chemical reactions that produce atmospheric constituents in excited states. These excited species then emit airglow photons as they return to lower energy states.

Mention will also be made here of the use of airglow limb scan profiles to obtain the exospheric temperature; measurement of molecular oxygen in absorption from airglow band ratios; and use of airglow ratios to obtain the O/N_2 ratio as a function of altitude, which is an important determinant of atmospheric energetics. Both nadir and limb airglow measurements can be used for obtaining composition.

The altitude profiles of several dayglow emissions can be used to extract the N_2, O_2, and O number densities and temperatures by matching the observations with models of airglow emission (*Meier and Anderson*, 1983,

15.3. AIRGLOW METHODS

Meier et al., 1985, and *Meier*, 1991). The determination of neutral thermospheric constituents in this manner is closely related to airglow methods for ionospheric electron density profiles, which are covered in Chapter 17.

DAY AIRGLOW: N_2

The ultraviolet day airglow has an abundance of nitrogen emission bands, (see Chapter 9). The most promising for molecular nitrogen composition measurements are the *Lyman-Birge-Hopfield* bands and the *Second Positive* bands. Other band systems, including the Vegard-Kaplan bands, the Birge-Hopfield bands, and the Caroll-Yoshino bands, either are weak or are difficult to relate simply to nitrogen number density.

The **Lyman-Birge-Hopfield** (LBH) bands of N_2, ($a\,^1\Pi_g - X\,^1\Sigma_g^+$), are found in the airglow in the approximately 120 to 220 nm region, with the strongest bands toward the lower end of this range. The nadir total band intensity is generally several kiloRayleigh, with the strongest individual bands up to 300 Rayleighs. The most useful for remote sensing are individual members of the series from (6,0) to (1,0) or broad band measurement of unresolved LBH emission in the 140 to 170 nm region.

Individual LBH bands and their suitability for remote sensing are as follows:

LBH (6,0) 127.3 nm Suitable for remote sensing, but not as strong as some other members.

LBH (5,0) 129.8 nm Not usable with current typical resolution of sensors, as very near the strong OI 130.4 nm dayglow emission.

LBH (4,0) 132.5 nm Excellent for remote sensing. Strong band separated from OI 130.4 and 135.6 nm dayglow emissions.

LBH (3,0) 135.4 nm Not usable as very near strong OI dayglow emission at 135.6 nm.

LBH (2,0), (5,2) 138.4 nm Excellent for remote sensing. Strong band readily resolved from OI.

LBH (4,2), (1,0) 141.6 nm Suitable for use as a single unresolved feature. There are other examples nearby.

LBH 140 to about 200 nm Many bands. Methods with broad filters useful. Longer wavelength limit of about 200 nm is due to interference by scattering.

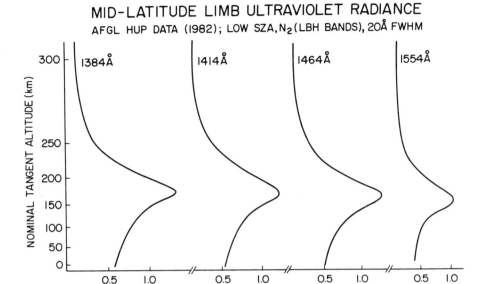

Figure 15.2: Limb profiles of four nitrogen LBH bands, HUP, *Huffman et al.*, 1981, 1983

The stronger individual bands in the list above that do not overlap atomic oxygen emission are at 127.3, 132.5, and 138.4 nm. Models of the limb profiles must account for solar zenith angle variations and solar activity levels. The emission decreases as the SZA increases, and the emission is much stronger at solar maximum than at solar minimum.

Figure 15.2 displays measurements of the nitrogen LBH airglow emission in the Earth limb. These were obtained from measurements on the shuttle by the HUP experiment, *Huffman et al.*, 1981, 1983. Figure 15.3 shows model calculations by *Meier*, 1991, for the 138.4 nm band for solar maximum and minimum and two solar zenith angles. Superimposed on these model calculations are data from a HUP measurement. Remote sensors using this method should be capable of measuring column emission rates from about three kiloRayleighs to near two Rayleighs.

The observed column emission is modified by absorption due to molecular oxygen in the Schumann-Runge continuum. This absorption can be used for obtaining oxygen number densities (*Meier and Anderson*, 1983).

The unfortunate overlap of several of the bands with other strong emis-

15.3. AIRGLOW METHODS

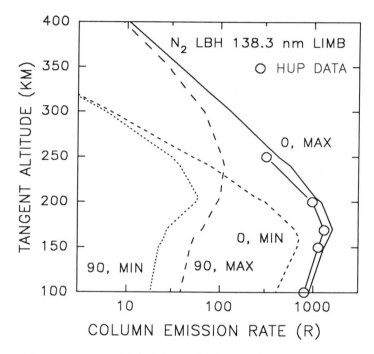

Figure 15.3: Nitrogen LBH 138.3 nm limb for solar max and min. Points: HUP, *Huffman et al.*, 1981, 1983. Model: *Meier*, 1991.

sion features is pointed out in the listing of the bands. The most troublesome is with the OI 135.6 nm emission, as discussed in connection with this emission. The strongest LBH bands are close enough in wavelength to other dayglow bands so that a spectrometer or spectrograph is required to separate them, given the difficulty of making interference filters in the FUV.

Use of the LBH bands should not be hampered by the discovery of spacecraft glow associated with this band system *Huffman et al.*, 1980, *Conway et al.*, 1987. This emission, which remains poorly understood, has only been observed at low altitudes (generally below 200 km) which is an unlikely altitude for a remote sensing satellite. The DMSP satellites, for example, are at an altitude of about 850 km, and the spacecraft glow decreases with the cube of the density. The LBH spacecraft glow is discussed in Chapter 5.

The N_2 **Second Positive** bands (2P), $C\,^3\Pi_u - B\,^3\Pi_g$, are another possibility for molecular nitrogen number density determination. This fully allowed transition is excited by energetic electron collision. The primary band utilized has been the (0,0) transition at 337.1 nm. The altitude distribution of the emission, obtained from sounding rocket or satellite limb scan measurements, has been used to measure molecular nitrogen concentration, photoelectron flux, and exospheric temperatures. It is possible that atmospheric temperatures will be obtainable from the rotational structure of the bands.

The intensity of the 337.1 nm band can reach over 25 kiloRayleighs in limb observations, and good filters are available, so that a spectrometer may not be needed. Atmospheric scattering prevents use in the nadir direction. Remote sensing with this emission, together with references to the earlier research, are discussed by *Meier and Anderson*, 1983, and by *Conway and Christensen*, 1985. If a limb viewing technique is used, this emission should be kept in mind for nitrogen composition, as it appears to be generally easier to use than the LBH bands.

DAY AIRGLOW: O

The 135.6 nm atomic oxygen emission, $2p^4\,^3P^o - 3s\,^5S$, is the airglow emission generally considered the best prospect for atomic oxygen composition measurements in the daytime. In contrast to the nearby and much more intense atomic oxygen resonance multiplet at 130.4 nm, the 135.6 nm emission is excited almost exclusively by photoelectron impact, and the contribution of solar fluorescence is small. In addition, multiple scattering is usually not a problem, but it must be included in modeling of limb profiles, *Strickland*

15.3. AIRGLOW METHODS

and Anderson, 1983. These properties useful in remote sensing applications result from the fact that 135.6 nm is not a fully allowed transition.

The use of 135.6 nm emission for atomic oxygen composition measurements and initial sounding rocket measurements are discussed by *Meier*, 1991. As laboratory and theoretical approaches improve, reanalysis of the sounding rocket data may be warranted, as demonstrated by *Link et al.*, 1988b.

Figure 15.4 presents limb scans based on calculations given by *Meier*, 1991, for solar maximum and minimum conditions. In addition, a few points from a measured limb scan based on Horizon Ultraviolet Program (HUP) data is shown, *Huffman et al.*, 1981, 1983. The calculations are done for the same conditions as the nitrogen LBH profiles in the last section. As is true for the nitrogen bands, both the number density and the exospheric temperature may be obtained from the limb scans. Intensities of between approximately 10 Rayleighs and 10 kiloRayleighs will be seen by a remote sensor, as shown in the figure.

There are difficulties in the use of 135.6 nm. First, it is located only about 5 nm from the 130.4 nm oxygen resonance line, which is 10 times brighter. A spectrometer is therefore required to separate the two, as interference filters are not currently available.

Another difficulty is with the presence of the strong (3,0) nitrogen LBH band at 135.4 nm, which is too close to the oxygen atom emission to be separated by most sensors flown to date. This band has an intensity about one-fifth of the oxygen emission. The approach generally used is to model the expected intensity of the LBH band and subtract this emission from the measured amount. Thus, it is an advantage to use a multiwavelength approach, in part to allow better LBH modeling.

Satellite composition experiments have been flown. The SSD (Secondary Sensor Density) experiment flown on the DMSP (Defense Meteorological Satellite Program) F4 satellite obtained limb emission profiles of several FUV and EUV wavelengths, *Pranke et al.*, 1982. An analysis of 135.6 nm limb profiles was reported by *Newman et al.*, 1983. A detailed analysis of the STP P78- 1 satellite measurements using revised experimental cross sections has been given by *Link et al.*, 1988a. The Imaging Spectroscopic Observatory, (ISO) has flown as a part of Spacelab on the shuttle, and it will fly on future missions. HUP measurements were mentioned previously. See Section 5.5 for further references to the space programs mentioned here. Although the results from these experiments are promising, many additional

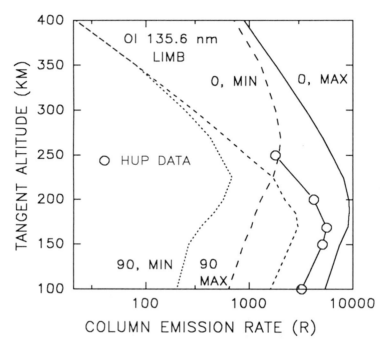

Figure 15.4: Atomic oxygen 135.6 nm limb. Model: *Meier*, 1991. Points: HUP, *Huffman et al.*, 1981, 1983.

measurements are needed to give the method the credibility required for widespread adoption.

NIGHT AIRGLOW: O AND N

The weak night airglow emissions from molecular oxygen in the Herzberg bands and nitric oxide in the delta and gamma bands can be used for remote sensing of atomic oxygen, O, and atomic nitrogen, N. The nightglow is due to chemiluminescent recombination reactions of these atoms, discussed in Chapter 9.

These emissions are relatively well understood and are always available to serve as methods for composition measurements at night. A possible difficulty is that individual bands are only a few Rayleighs each. However, integration over time from a moving satellite is usually acceptable. The relationship of atomic oxygen to the UV nightglow and to the 557.7 nm airglow line has been recently discussed by *Siskind and Sharp*, 1991.

15.4 Recommended methods

This section lists what appear to be the best methods at this time for ultraviolet remote sensing of each neutral constituent of the mesosphere and thermosphere. These methods are described more fully elsewhere in this chapter.

Molecular oxygen, O_2 Absorption in the Schumann-Runge continua and bands offers a wide range of cross section values for occultation use. Differential absorption by LBH bands scanned in the limb is feasible.

Molecular nitrogen, N_2 Airglow from the LBH bands can be used with both nadir and limb viewing.

Atomic oxygen, O Airglow from the 135.6 nm emission line is the method of choice for both nadir and limb viewing in the daytime. The Herzberg band emission can be used at night.

Nitric oxide, NO Fluorescence from the gamma bands observed in the limb in the daytime is the preferred method.

Atomic hydrogen, H Fluorescence scattering at Lyman alpha (121.6 nm) can be utilized.

Atomic helium, He Fluroescence scattering at the 58.4 nm resonance line is available.

Atomic nitrogen, N The delta and gamma bands of the nightglow can be used.

These methods can be combined in models and algorithms that will obtain several constituents at once. In this case, the recommended method might change. For example, where it can be used, the **EUV absorption method** for O_2, N_2, and O appears to be an excellent choice. Some of these methods will likely be incorporated into global space weather systems, as discussed in Chapter 17.

15.5 Atmospheric density

Satellite and space debris orbit prediction becomes increasingly important as the use of space increases. The need is for improved total density predictions, so that satellite drag can be more accurately modeled. Standard atmospheres, MSIS models, and general circulation models, briefly discussed in Chapter 6, are not yet capable of providing densities to the accuracies needed. Improvement of density modeling may well depend on improved remote sensing methods.

The satellite orbit prediction problem is most acute in the upper part of the mesosphere, and it has been the subject of investigation for a number of years, (see *Champion*, 1983, and references therein). At these altitudes, satellites are subjected to so much atmospheric drag that an orbiting vehicle does not spend much time there. Thus, the in-situ accelerometer technique, *Champion and Marcos*, 1973, so important in developing predictive models at higher altitudes, is not effective at the altitudes where the density measurement is most important for satellite reentry prediction. Other methods of density measurement are needed.

The current accuracy of satellite drag models was recently discussed by *Marcos*, 1990, and the need for improved density measurements was noted. Accuracies in the 5 to 10% range are sought. Remote sensing methods such as those given in this chapter could provide these density measurements.

Another approach to density and composition sensing is to improve our knowledge of **solar EUV and FUV radiation**, which is becoming recognized as one of the principal unknowns in developing atmospheric monitoring methods. The combination of solar monitoring with global circulation models may lead to significant improvements in density prediction. The strong influence of solar emission on the density and thus on satellite drag over the solar cycle is well known, as was recently reviewed by *Gourney*, 1990, and *Walterscheid*, 1989.

A feasibility study for an atmospheric density satellite, *Killeen*, 1989, includes solar EUV measurements as a useful enhancement to the more traditional methods. The flux for the separate solar lines, which vary individually over the solar cycle, is wanted. It should be pointed out that not only the EUV flux is needed for these models. The absorption by oxygen in the Schumann-Runge continuum and bands is very important as well, and so accurate FUV flux values and transmission models are also needed. Solar flux measurements are discussed in Chapter 7. The use of solar flux in global models of ionospheric electron density is discussed in 17.3.

15.6 References

Anderson, D. E., Jr., R. R. Meier, and C. S. Weller, The seasonal-latitudinal variation of exospheric helium from He 584 Å dayglow emissions, *J. Geophys. Res.*, *84*, 1914, 1979.

Anderson, D. E., Jr., L. J. Paxton, R. P. McCoy, and R. R. Meier, Atomic hydrogen and solar Lyman alpha flux deduced from STP 78-1 UV observations, *J. Geophys. Res.*, *92*, 8759–8766, 1987.

Anderson, G. P., H. S. Muench, R. E. Good, C. R. Philbrick, and W. Swider, Atmospheric Composition, Chapter 21 in *Handbook of Geophysics and the Space Environment*, edited by A. S. Jursa, pp. 21-1 to 21-67, Air Force Geophysics Laboratory, NTIS ADA 167000, 1985.

Atreya, S. K., T. M. Donahue, W. E. Sharp, B. Wasser, J. F. Drake, and G. R. Riegler, Ultraviolet stellar occultation measurement of the H_2 and O_2 densities near 100 km in the earth's atmosphere, *Geophys. Res. Lett.*, *3*, 607–610, 1976.

Barth, C. A., Rocket measurement of the nitric oxide dayglow, *J. Geophys. Res.*, *69*, 3301, 1964.

Barth, C. A., Nitric oxide in the upper atmosphere, *Ann. Geophys.*, *22*, 198, 1966.

Barth, C. A., Reference models for thermospheric NO, *Adv. Space Sci.*, *10*, 103–116, 1990.

Barth, C. A., D. W. Rusch, R. D. Thomas, G. H. Mount, G. J. Rottman, G. E. Thomas, R. W. Sanders, and G. M. Lawrence, Solar Mesospheric Explorer: Scientific objectives and results, *Geophys. Res. Lett.*, *10*, 237–240, 1983.

Barth, C. A., W. K. Tobiska, D. E. Siskind, and D. D. Cleary, Solar-terrestrial coupling: Low-latitude thermospheric nitric oxide, *Geophys. Res. Lett.*, *15*, 92, 1988.

Bowers, C. W., P. D. Feldman, P. D. Tennyson, and M. Kane, Observation of the OI ultraviolet intercombination emissions in the terrestrial dayglow, *J. Geophys. Res.*, *92*, 239–245, 1987.

Bruner, Jr., E. C. and T. E. Wilson, Deuterium in the Earth's exosphere, *J. Geophys. Res.*, *74*, 6491–6493, 1969.

Chakrabarti, S., F. Paresce, S. Bowyer, R. Kimble, and S. Kumar, The extreme ultraviolet day airglow, *J. Geophys. Res.*, *88*, 4898, 1983.

Champion, K. S. W., Properties of the mesosphere and thermosphere and comparison with CIRA 72, *Adv. Space Res.*, *3*, 45, 1983.

Champion, K. S. W. and F. A. Marcos, The triaxial accelerometer system on Atmospheric Explorer, *Radio Sci.*, *8*, 197, 1973.

Cleary, D. D., Daytime high-latitude rocket observation on the NO γ, δ, and ϵ bands, *J. Geophys. Res.*, *91*, 11,337, 1986.

Cleary, D. D., R. R. Meier, E. P. Gentieu, P. D. Feldman, and A. B. Christensen, An analysis of the effects of N_2 absorption on the O^+ 834 Å emission from rocket observations, *J. Geophys. Res.*, *94*, 17,281, 1989.

Conway, R. R., Self-absorption of the N_2 Lyman-Birge-Hopfield bands in the far ultraviolet dayglow, *J. Geophys. Res.*, *87*, 859, 1982.

Conway, R. R., Multiple fluorescent scattering of N_2 ultraviolet emissions in the atmospheres of the Earth and Titan, *J. Geophys. Res.*, *88*, 4784, 1983.

Conway, R. R. and A. B. Christensen, The ultraviolet dayglow at solar maximum 2. Photometer observations of N_2 second positive (0,0) band emission, *J. Geophys. Res.*, *90*, 6601–6607, 1985.

Conway, R. R., R. R. Meier, and R. E. Huffman, The far ultraviolet vehicle glow on the S3-4 satellite, *Geophys. Res. Lett.*, *14*, 628, 1987.

Conway, R. R., R. R. Meier, and R. E. Huffman, Abundance of atomic oxygen in the lower thermosphere from satellite observations of the OI 164.1 Å dayglow, *Planet. Space Sci.*, *36*, 963, 1988.

Cravens, T. E., Nitric oxide gamma band emission rate factor, *Planet. Space Sci.*, *25*, 369, 1977.

Cravens, T. E., J.-C. Gérard, M. LeCompte, A. I. Stewart, and D. W. Rusch, The global distribution of nitric oxide in the thermosphere as determined by the Atmosphere Explorer D satellite, *J. Geophys. Res.*, *90*, 9862–9870, 1985.

Erdman, P. W. and E. C. Zipf, Remote sensing of atomic oxygen: Some observational difficulties in the use of the forbidden OI λ1173 Å and OI λ1641 Å transitions, *J. Geophys. Res.*, *92*, 10,140–10,144, 1987.

Fahr, H. J. and T. Smid, Spectrophotometric EUV observations and the theoretical modeling of the geocoronal HeI 584/537 Å radiation field, *Ann. Geophys.*, *4*, 447, 1986.

Friedman, H., S. W. Lichtman, and E. T. Byram, Photon counter measurements of solar x-rays and extreme ultraviolet light, *Phys. Rev.*, *83*, 1025–1030, 1951.

Garriott, O. K., R. B. Norton, and J. G. Timothy, Molecular oxygen concentrations and absorption cross sections in the thermosphere derived from extreme ultraviolet occultation profiles, *J. Geophys. Res.*, *82*, 4973–4982, 1977.

Gladstone, G. R., UV resonance line dayglow emissions on Earth and Jupiter, *J. Geophys. Res.*, *93*, 14,623–14,630, 1988.

Gladstone, G. R., R. Link, S. Chakrabarti, and J. C. McConnell, Modeling of the OI 989 Å to 1173 Å ratio in the terrestrial dayglow, *J. Geophys. Res.*, *92*, 12,445–12450, 1987.

Gourney, D. J., Solar cycle effects on the near-earth space environment, *Rev. Geophys.*, *28*, 315–336, 1990.

Green, A. E. S., *The Middle Ultraviolet: Its Science and Technology*, A. E. S. Green, editor, p. 127, Wiley, 1966.

Grossman, K. U., K. S. W. Champion, M. Roemer, W. L. Oliver, and T. A. Blix, editors, *The Earth's Middle and Upper Atmosphere*, Adv. Space Res., *7*, No. 10, 1987, and Pergamon Press, 1987.

Hall, L. A., W. Schweizer, and H. E. Hinteregger, Diurnal variation of atmosphere around 100 km from solar EUV absorption measurements, *J. Geophys. Res.*, *68*, 6413–6417, 1963.

Hall, L. A., W. Schweizer, and H. E. Hinteregger, Improved extreme ultraviolet absorption measurements in the upper atmosphere, *J. Geophys. Res.*, *70*, 105–111, 1965.

Hays, P. B. and R. G. Roble, Stellar spectra and atmospheric composition, *J. Atmos. Sci.*, *25*, 1141–1153, 1968a.

Hays, P. B. and R. G. Roble, Atmospheric properties from the inversion of planetary occultation data, *Planet. Spa. Sci.*, *16*, 1197–1198, 1968b.

Hays, P. B. and R. G. Roble, Stellar occultation measurements of molecular oxygen in the lower thermosphere, *Planet. Spa. Sci.*, *21*, 339–348, 1972.

Hays, P. B., R. G. Roble, and A. N. Shah, Terrestrial atmospheric composition from stellar occultations, *Science*, *176*, 793–794, 1972.

Hinteregger, H. E., Absorption spectrometric analysis of the upper atmosphere in the EUV region, *J. Atmos. Sci.*, *19*, 351–368, 1962.

Hinteregger, H. E. and L. A. Hall, Thermospheric densities and temperatures from EUV absorption measurements by OSO-III, *Space Res.*, *9*, 519–529, 1969.

Huffman, R. E. and J. C. Larrabee, Effect of absorption by atomic oxygen and atomic nitrogen lines on upper atmosphere composition measurements, *J. Geophys. Res.*, *73*, 7419–7428, 1968.

15.6. REFERENCES

Huffman, R. E., F. J. LeBlanc, D. E. Paulsen, and J. C. Larrabee, Satellite vacuum ultraviolet airglow and auroral observations, *J. Geophys. Res.*, *85*, 2201–2215, 1980.

Huffman, R. E., F. J. LeBlanc, D. E. Paulsen, and J. C. Larrabee, Ultraviolet horizon sensing from space, in *Shuttle Pointing of Electro-optical Experiments*, W. Jerkovsky, editor, *Proc. SPIE 265*, 290–294, 1981.

Huffman, R. E., F. J. LeBlanc, J. C. Larrabee, D. E. Paulsen, and V. C. Baisley, Horizon radiance measurements from shuttle, *AIAA Shuttle Environment and Operations Meeting, Washington, D. C.*, Paper AIAA-83-262B-CP, Nov., 1983.

Jones, R. A., E. C. Bruner, Jr., and W. A. Rense, Absorption measurements of Earth's hydrogen atmosphere from solar hydrogen Lyman alpha rocket data, *J. Geophys. Res.*, *75*, 1849–1853, 1970.

Jursa, A. S., Y. Tanaka, and F. J. LeBlanc, Nitric oxide and molecular oxygen in the earth's upper atmosphere, *Planet. Spa. Sci.*, *1*, 161–172, 1959.

Jursa, A. S., M. Nakamura, and Y. Tanaka, Molecular oxygen distribution in the upper atmosphere, *J. Geophys. Res.*, *68*, 6145–6155, 1963.

Jursa, A. S., M. Nakamura, and Y. Tanaka, Molecular oxygen distribution in the upper atmosphere, 2, *J. Geophys. Res.*, *70*, 2699–2702, 1965.

Killeen, T. L., A. G. Burns, B. C. Kennedy, R. G. Roble, and F. A. Marcos, Feasibility study for an atmospheric density specification satellite, *GL-TR-89-0081, Environmental Research Papers No. 1026*, Air Force Geophysics Laboratory, 1989.

Knight, D. E., R. Uribe, and B. E. Woodgate, Extreme ultraviolet absorption cross sections in the earth's upper atmosphere, *Planet. Spa. Sci.*, *20*, 161–164, 1972.

Link, R., S. Chakrabarti, G. R. Gladstone, and J. C. McConnell, An analysis of satellite observations of the OI EUV dayglow, *J. Geophys. Res.*, *93*, 2693–2714, 1988a.

Link, R., G. R. Gladstone, S. Chakrabarti, and J. C. McConnell, A reanalysis of rocket measurements of the ultraviolet dayglow, *J. Geophys. Res.*, *93*, 14,631–14,648, 1988b.

Marcos, F. A., Accuracy of atmospheric drag models at low satellite altitudes, *Adv. Space Res.*, *10*, (3)417–(3)422, 1990.

Meier, R. R., Ultraviolet spectroscopy and remote sensing of the upper atmosphere, *Space Sci. Rev.*, *58*, 1–185, 1991.

Meier, R. R. and C. S. Weller, EUV resonance radiation from helium atoms and ions in the geocorona, *J. Geophys. Res.*, *77*, 1190, 1972.

Meier, R. R. and D. E. Anderson, Jr., Determination of atmospheric composition and temperature from the UV dayglow, *Planet. Space Sci.*, *31*, 967, 1983.

Meier, R. R. and R. R. Conway, The $^1D - {^1}S$ transition in atomic oxygen: A new method of measuring the O abundance in planetary thermospheres, *Geophys. Res. Lett.*, *12*, 601, 1985.

Meier, R. R., R. R. Conway, D. E. Anderson, Jr., P. D. Feldman, R. W. Eastes, E. P. Gentieu, and A. B. Christensen, The ultraviolet dayglow at solar maximum 3. Photoelectron excited emissions of N_2 and O, *J. Geophys. Res.*, *90*, 6608–6616, 1985.

Newman, A. L., A. B. Christensen, and D. E. Anderson, Jr., Calculated and observed limb profiles of OI (1356 Å) dayglow, *J. Geophys. Res.*, *88*, 9265–9270, 1983.

Nicolet, M., Contribution à l'étude de la structure de l'ionosphère, *Inst. Roy. Meteorol. Belg. Mem.*, *19*, 162, 1945.

Norton, R. B. and J. M. Warnock, Seasonal variation of molecular oxygen near 100 kilometers, *J. Geophys. Res.*, *73*, 5798–5800, 1968.

Opal, C. B. and H. W. Moos, Nighttime molecular oxygen densities in the 100 to 130 km region from Schumann-Runge absorption, *J. Geophys. Res.*, *74*, 2398–2401, 1969.

Pranke, J. B., A. B. Christensen, F. A. Morse, D. R. Hickman, W. T. Chater, C. K. Howey, and D. A. Jones, A satellite borne limb scanning ultraviolet spectrometer for thermospheric remote sensing, *Appl. Opt.*, *21*, 3941–3952, 1982.

Prinz, D. K. and G. E. Brueckner, Observations of the O_2 column density between 120 km and 70 km and absorption cross section in the vicinity of H-Lyman alpha, *J. Geophys. Res.*, *82*, 1481, 1977.

15.6. REFERENCES

Rairden, R. L., L. A. Frank, and J. D. Craven, Geocoronal imaging with Dynamics Explorer, *J. Geophys. Res., 91*, 13,613–13,630, 1986.

Rees, M. H., *Physics and Chemistry of the Upper Atmosphere*, Cambridge, 1989.

Rusch, D. W., Satellite ultraviolet measurements of nitric oxide fluorescence with a diffusive transport model, *J. Geophys. Res., 78*, 5676–5686, 1973.

Rusch, D. W., G. H. Mount, C. A. Barth, R. J. Thomas, and M. T. Callan, Solar mesosphere explorer spectrometer: Measurements of ozone in the 1.0 to 0.1 mbar region, *J. Geophys. Res., 89*, 11,677, 1984.

Sharp. W. E., The measurement of atomic oxygen in the mesosphere and lower thermosphere, *Planet. Space Sci., 39*, 617–626, 1991.

Singer, S. F., Geophysical research with artificial earth satellites, in *Advances in Geophysics, 3*, H. E. Landsberg, editor, 301–367, Academic Press, 1956.

Siskind, D. E. and W. E. Sharp, A comparison of measurements of the oxygen nightglow and atomic oxygen in the lower thermosphere, *Planet. Space Sci., 39*, 627–639, 1991.

Siskind, D. E., C. A. Barth, and R. G. Roble, The response of thermospheric nitric oxide to an auroral storm, 1, Low and middle latitudes, *J. Geophys. Res., 94*, 16,885–16,898, 1989a.

Siskind, D. E., C. A. Barth, D. S. Evans, and R. G. Roble, The response of thermospheric nitric oxide to an auroral storm, 2, Auroral latitudes, *J. Geophys. Res., 94*, 16,899–16,911, 1989b.

Smith, G. R. and D. M. Hunten, Study of planetary atmospheres by absorptive occultations, *Rev. Geophysics, 28*, 117–143, 1990.

Strickland, D. J. and D. E. Anderson, Jr., Radiation transport effects on the OI 1356 Å limb intensity profile in the dayglow, *J. Geophys. Res., 88*, 9260–9264, 1983.

Tisone, G. C., Measurements of the absorption of solar radiation in the region of 2150 Å by O_2 and O_3, *J. Geophys. Res., 77*, 2971, 1972.

Torr, D. G., The photochemistry of the upper atmosphere, Chapter 5 in *The Photochemistry of Atmospheres: Earth, the Other Planets, and Comets*, J. S. Levine, editor, 165–278, Academic, 1985.

Walterscheid, R. L., Solar cycle effects on the upper atmosphere: Implications for satellite drag, *J. Spacecraft and Rockets*, *26*, 439, 1989.

Weeks, L. H., Determination of O_2 density from Lyman α ion chambers, *J. Geophys. Res.*, *80*, 3655–3660, 1975a.

Weeks, L. H., Observation of O_2 variability at mid-latitudes from 1450 Å measurements, *J. Geophys. Res.*, *80*, 3661–3666, 1975b.

Weeks, L. H. and L. G. Smith, Molecular oxygen concentrations in the upper atmosphere by absorption spectroscopy, *J. Geophys. Res.*, *73*, 4835–4849, 1968.

Chapter 16

Global Auroral Imaging

We come now to imaging of the aurora from space using its naturally occurring ultraviolet emissions. This major application area, demonstrated during in the last decade, has led to adoption of UV imaging as an important component in **solar-terrestrial research programs**. Global UV auroral imaging, combined with other sensors, will also be incorporated into **global space weather systems** that will improve the operation of communications, navigation, and radar systems, as discussed in the next chapter.

This chapter and the next are closely related. For both, use is primarily made of the FUV (about 200 to 100 nm) and the EUV (about 100 to 10 nm).

In the present chapter, we describe UV auroral imaging in more detail, including a history of the development of imagers, technical descriptions of the imagers flown to date, the relation to other auroral sensing techniques, and promising research applications. An introduction to the aurora is given in Chapter 10.

16.1 The aurora from space

The availability of orbiting satellites as platforms for imaging large regions of the Earth and its atmosphere from space has led to many applications. Imagery of the land and ocean surfaces with LANDSAT, SPOT, and various military satellites is now routinely done for applications as diverse as Earth resource utilization and defense surveillance. Atmospheric, as opposed to surface, uses have included applications to meteorology with satellite programs such as GOES, NIMBUS, and DMSP, which obtain images of clouds and other measurements to provide real-time weather information and to aid in forecasting.

We discuss here satellite observations of the thermospheric regions of the atmosphere, where the naturally occurring aurora, airglow, and fluorescence emission sources can be monitored with UV sensors. Emphasis in this chapter is on nadir and Earth viewing directions, as used by auroral imagers, rather than limb viewing.

Satellite UV detection of the aurora was demonstrated in early space observations. Satellite nadir-viewing photometers having broad-band coverage in the FUV clearly demonstrated that the aurora could be observed from space, above the atmospheric background and absorption, as reported by *Joki and Evans*, 1969, from data collected in 1965, and by *Chubb and Hicks*, 1970, using OGO-4 satellite data (1968). The sensors obtained nadir radiance measurements along the ground track of the satellite, with a peak observed at the location of the auroral oval.

Rocket spectra of the aurora were being acquired a few years before satellite measurements, as outlined in Chapter 10. However, few nadir measurements were available, since the sounding rockets usually did not get to the altitudes of greater than 200 km needed to be above most of the auroral emission, and many flights did not have control of sensor pointing.

The first global UV images and spectra of the entire Earth and its atmosphere were obtained from a camera on the moon during the Apollo program, *Carruthers and Page*, 1972, 1976a,b; *Carruthers et al.*, 1976. These dramatic images showed auroral ovals, equatorial airglow belts, and dayside airglow in images obtained using a broad FUV band that excluded H Lyman alpha. The rather structureless geocorona was imaged in separate frames taken in H Lyman alpha emission. These images were and are extremely influential in demonstrating the potential of UV global imaging. One is displayed as Figure 2.3.

Visible wavelength auroral imagery was obtained by the ISIS-II satellite *Anger et al.*, 1973, *Lui et al.*, 1975. At about the same time, *Rogers et al.*, 1974, reported visible region images of aurora from DOD meteorological satellites. The latter series of satellites, currently referred to as the Defense Meteorological Satellite Program, or DMSP, has been extremely important in auroral studies. The visual region imagers, used for cloud images during the daytime, provide auroral images at night. These images are obtained as a by-product of the cloud images on the operational satellite system, resulting in a tremendous number of images. A number of auroral studies continue to be reported using these images and other sensors carried by these satellites, with a review of initial work by *Akasofu*, 1976.

Auroral UV spectra from OGO-4 were reported by *Gerard and Barth*, 1976. Subsequent experiments were at higher sensitivities and with improved spectral resolution. They provided, among other results, intensities useful in the design of future imagers. These measurements were reported by *Huffman et al.*, 1980, from the S3-4 satellite flown in 1978, and by *Paresce et al.*, 1983, from the P78-1 satellite flown in 1979. The nadir viewing UV traces across the auroral oval from the S3-4 satellite were shown to be very similar to DMSP visible images taken at the same time, *Huffman et al.*, 1981b.

The DMSP and ISIS-II visible images created interest in the study of the aurora using global images from satellites. With the recognition that there was sufficient intensity in the UV for imaging with reasonably sized sensors, the use of UV imagers could be seriously proposed. All the elements needed for successful development of satellite UV imagers were available.

16.2 Satellite UV imagers

Satellite imagers flown primarily to observe the aurora and other features of the atmosphere and having UV sensors are described in this section. Note that over time the imagers tend to be referred to by the satellite name, rather than by the sensor or experiment name. This practice may occasionally cause some confusion. In addition to the discussion in this section, these programs may be mentioned in other chapters in connection with the use of the data both for global imaging and for other remote sensing applications.

EXOS-A (KYOKKO) The first experiment dedicated to imaging the aurora in the FUV was reported by *Hirao and Itoh*, 1978, and by *Kaneda*, 1979. They used a modified television camera sensitive to the hydrogen Lyman alpha line emission at 121.6 nm on the Japanese satellite KYOKKO, which is also called EXOS-A. While this emission is produced by proton precipitation in the auroral zone, it may be difficult to observe due to the intense geocoronal emission (see Chapter 11).

The television camera used as the imager has been described by *Kaneda et al.*, 1977. Images from this flight have been published by *Kaneda et al.*, 1981, but there is apparently not a large data base.

DYNAMICS EXPLORER The next imager experiment to include

UV was flown as a part of the NASA Dynamics Explorer program. Two satellites, called DE-1 and DE-2, were placed in orbit by a single booster on August 3, 1981. The DE-1 satellite carrying the imagers continued to make observations until early 1991. Possibly the best introduction to the program has been given in a series of papers in *Reviews of Geophysics* and introduced by *Hoffman*, 1988, after five years of operation of these satellites. Both satellites had a number of instruments. The imager experiment was on DE-1, which was in a polar orbit at $90°$ inclination with perigee and apogee altitudes of 570 km and 3.65 Earth radii (23,000 km) and an orbital period of 6.83 hours. The orbit precesses around the Earth at a rate of $0.328°$ per day, so that the part of the orbit closest to the apogee, where the best images of the total Earth are obtained, is directly above a polar region about every 1.5 years.

The auroral imaging instrument on DE-1 was called the Spin Scan Auroral Imager. The principal investigator was L. A. Frank, with the scientific collaboration of J. D. Craven and others. The principal institution is the University of Iowa. The auroral imaging instrumentation is described by *Frank et al.*, 1981. There are three spin-scan imagers, two in the visible region and one for the UV, utilizing photomultiplier tubes as the light measuring element. Each has a filter wheel allowing one of twelve filters in the light path. The visible imagers include filters to obtain the atomic oxygen airglow "red line" at 630.0 nm and "green line" images at 557.7 nm; the N_2^+ (0,0) first negative band images at 391.4 nm; ozone column densities using filters at 317.5 and 360.0 nm; an exploratory search for marine bioluminescence at 482.5 nm; and a cloud distribution experiment utilizing a filter at 629.8 nm.

The ultraviolet imager had filters transmitting the atomic oxygen multiplets 130.4 and 135.6 nm, hydrogen Lyman alpha at 121.6 nm, and the molecular nitrogen Lyman-Birge-Hopfield (LBH) bands between about 140 and 170 nm. The filters used had bandwidths sufficiently wide so that data from several filters must be combined to cleanly separate the oxygen multiplet and the nitrogen LBH emission. The imager experiment is designed with particular attention to rejecting scattered light from the sunlit parts of the Earth image, so that airglow in the dark side can be observed. This objective was successfully accomplished using superpolished optics and other techniques.

Imaging is accomplished in the following manner. The satellite is spin-stabilized at 10 revolutions per minute, with the spin axis perpendicular to the orbit plane. The photometers have fields of view toward the Earth that

16.2. SATELLITE UV IMAGERS

can be deflected about an axis perpendicular to the spin axis by an internal rotating mirror. Each scan line, acquired in six seconds, provides a strip image across the Earth. Every revolution, the rotating mirror moves $0.25°$, thus moving the position of the strip so that it moves across the Earth direction. After 120 rotations of the spacecraft, requiring 720 seconds, a typical $30°$ wide image is obtained, which near apogee is sufficient to image the entire Earth. The instantaneous field of view is $0.32°$, resulting in a spatial extent at the aurora from apogee of about 130 km.

An initial report on auroral images was given by *Frank et al.*, 1982. In contrast to the other satellites flown to date, data is also obtained at low latitudes, including images of the equatorial airglow belts as well as the auroral oval images. At this point, there have been a number of studies that have used the Dynamics Explorer data base of UV and other images, which has several hundred thousand images. This data base is available to qualified users either through the University of Iowa or through the National Space Science Data Center, Goddard SFC. Summary scientific articles have been published by *Frank et al.*, 1985, and *Frank and Craven*, 1988.

HILAT and POLAR BEAR The HILAT (HIgh LATitude) satellite launched in 1983 and the Polar BEAR (Polar Beacon Experiment and Auroral Research) satellite launched in 1986 carried similar auroral imagers in circular, polar orbits at altitudes of about 830 and 1000 km, respectively. These altitudes are much lower than the DE-1 and Viking altitudes but are still sufficiently high to image almost all of the average auroral oval found when $K_p = 3$. These two similar spacecraft were stabilized in all three axes; had no on-board data storage, limiting coverage to dedicated receiving stations; and were primarily operated when the satellites were over the northern polar region. One of the goals of these DOD (Department of Defense) satellites was to investigate radiowave scintillation effects caused by ionospheric irregularities, by means of radiowave transmission data from a satellite beacon, as described further in Section 17.4.

The HILAT program was described initially by *Fremouw et al.*, 1983, 1985. These programs have been described in more detail in two series of papers (HILAT, *Potemra*, 1984; Polar BEAR, *Grant*, 1987). The introductory paper for HILAT, by the program scientist and representative of the principal government sponsor, respectively, for both programs, *Fremouw and Wittwer*, 1984, serves as a scientific introduction for both satellite programs. The principal investigators for the imagers on both satellites were

258   CHAPTER 16. GLOBAL AURORAL IMAGING

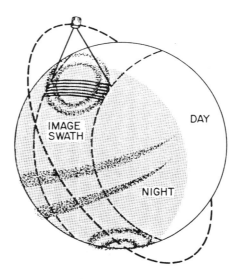

Figure 16.1: Global imaging from the HILAT and Polar BEAR satellites

R. E. Huffman, Air Force Geophysics Laboratory (now Phillips Laboratory, Geophysics Directorate) and C.-I. Meng, Applied Physics Laboratory.

The HILAT imager is called AIM, for Auroral Ionospheric Mapper, while the Polar BEAR imager is called AIRS, for Auroral Ionospheric Remote Sensor. Both imagers are mounted on the underside of the three-axis stabilized satellite. For both AIM and AIRS, a scanning mirror is used to view a strip of atmospheric radiance from slightly above the horizon, through nadir, to the other horizon in the direction perpendicular to the satellite ground track. The scan requires a total time of three seconds, during which time the satellite has moved in its orbit by about 20 km, so that the next cycle of the scanning mirror covers the adjoining region.

As the satellite continues to move in a roughly north-south direction, the scanning mirror observes adjoining strips in a roughly east-west direction. When the single stream of bits from counting periods of about seven milliseconds is placed in correct registration on the ground and intensity ranges are displayed in false colors, an image of the aurora above the airglow is obtained. The general concept of the imaging used by these satellites is shown in Figure 16.1.

The imaging approach used here is similar to the Operational Line Scan visible wavelength imager on the DMSP satellites, which is designed for cloud

16.2. SATELLITE UV IMAGERS

images in the daytime and used for auroral imaging at night. AIM and AIRS were made generally compatible with the DMSP spacecraft in order to serve as proof of concept demonstrations for future operational imagers. At this time, it is planned to fly similar operational UV imagers on the DMSP spacecraft in the late 1990s.

Note that this method of imaging obtains one auroral image for each pass, and thus it cannot observe dynamic motions that would be seen in a series of snapshot mode images. The time between successive images of the aurora is the period of the orbit, which was about 90 minutes for HILAT and 100 minutes for Polar BEAR. The spatial resolution however can be somewhat better, with the footprint of the FOV at the aurora being about 5 by 20 km. The footprint becomes much larger as the scan moves to the Earth limb. The lower altitude operational meteorological satellites such as DMSP and NIMBUS are usually in a sun-synchronous orbit, with an inclination of about $98.6°$ so that the local times are the same on each orbit. In order to study solar zenith angle effects, HILAT and Polar BEAR were not in sun-synchronous orbits, and the local time changed gradually throughout the missions.

These imagers differ from the DE-1 and VIKING imagers by using a spectrometer to isolate wavelength bands of interest in the FUV rather than filters. Thus, it is also possible to obtain spectra of the aurora and airglow. The sensors can operate in imaging mode, with the spectrometer scan fixed at a commanded position; spectrometer mode, with the scan mirror locked in the nadir direction while the spectrometer scans a range from roughly 110 to 180 nm; and the photometer mode, with the scan mirror locked in the nadir direction and the spectrometer set at a fixed wavelength position.

The AIM imager on HILAT uses a one-eighth meter Ebert Fastie spectrometer with an EMR 510 photomultiplier having a cesium iodide photocathode and a MgF_2 window as the single detector. The field of view, determined by the entrance slit of the spectrometer, is $1.53°$ along track and $0.373°$ cross track. The total angle scanned cross track is $134.4°$, which is divided into 336 data samples corresponding to pixels by using steps and corresponding counting periods of 6.8 milliseconds. The spectrometer is placed at the focus of a reflecting telescope illuminated by the scanning mirror. The spectrometer, photomultiplier, and telescope used on AIM are based on the sensor used for the Horizon Ultraviolet Program (HUP) on shuttle flights STS-4 and STS-39, as described by *Huffman et al.*, 1981a. A discussion of the concept and sensor is given by *Meng et al.*, 1983. The AIM sensor as flown is described by *Schenkel and Ogorzalek*, 1984, and *Schenkel et al.*, 1985.

The spectrometer has a FWHM bandwith of 3.0 nm. A series of fifteen commandable wavelengths are possible for the imaging mode from hydrogen Lyman alpha at 121.6 nm to a nitrogen Lyman-Birge-Hopfield (LBH) band at 175.1 nm. These bands are centered on important auroral emitters including also atomic oxygen emission at 130.4 nm and 135.6 nm, atomic nitrogen emission at 149.3 nm, and molecular nitrogen LBH band peaks at 141.5, 145.4, 159.5, 167.0, and 175.1 nm. The spectral bandwidth of 3.0 nm is not sufficient to isolate single LBH bands.

The satellite was launched on June 27, 1983, into a near-circular orbit at about 830 km with an inclination of about 83°. Unfortunately, the imager only operated for about one month before an apparent electronics failure occurred. It nevertheless provided convincing evidence that this method of imaging the aurora had promise for both day and night imaging, and it led to improvements in the design of AIRS.

The AIRS sensor on the Polar BEAR satellite is a continuation of the approach used in AIM but with the provision for two channels of imaging in the FUV and for two additional images selected from NUV, MUV, and visible filters. A schematic layout of the sensor is shown in Figure 16.2. The multiple imaging capability in the design is both for redundancy and for improved auroral image analysis using multiple bands. It represents a transition between sensors that use array detectors at the focal plane of spectrometers and sensors utilizing photomultiplier tubes to scan through the spectrum. However, the design is suitable for use as an operational sensor for global space weather systems.

For AIRS, the spectrometer was modified so that there are two exit slits and two photomultipliers. The two slits were placed 24 nm apart in the spectrum, after consideration of useful pairs of emission bands. The focal length of the exit beam had to be increased to allow enough room for two EMR 510 photomultipliers to be installed behind the exit slits. Additional images were obtained by using a small hole in the center of the reflecting telescope to obtain a second beam from the scanning mirror. This beam was split and passed through one of two pairs of fixed filters and then to two photomultipliers. In this manner, four simultaneous images at four wavelengths can be obtained.

The FUV spectrometer covers any wavelength between 115 and 180 nm for either channel. It is possible to command pairs of images separated by

16.2. SATELLITE UV IMAGERS

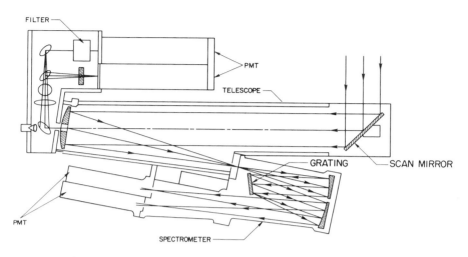

Figure 16.2: AIRS ultraviolet imager optical schematic

24 nm using the most prominent spectral features, with the most used being the oxygen atom 135.6 nm emission and a group of nitrogen LBH bands centered at 159.6 nm. The spectral bandwidth in both channels was 3.6 nm. The two additional images could be obtained using either a pair of filters at 630 nm, oxygen red line and 391.4 nm, N_2^+ (0,0) first negative band, or a pair covering the 337.1 nm N_2 (0,0) second positive band and a broad MUV scatter band centered on 225 nm. The geometrical relationships of the orbit, field of view, and spatial coverage are shown in Figure 16.3.

The remaining specifications, operations constraints, and philopshy of AIRS are identical or very similar to the earlier AIM sensor. AIRS instrumentation and use has been described by *Schenkel et al.*, 1986 and *Schenkel and Ogorzalek*, 1987.

The Polar BEAR satellite was launched on November 13, 1986, and the imager acquired useful data over a period of about two years and four months to May 1989. The data base contains about 6000 images. Coverage during this period is not complete, due to attitude control problems with the satellite and some experimental anomalies.

Scientific descriptions of the auroral images from the HILAT satellite have been reported by *Meng and Huffman*, 1984; *Huffman and Meng*, 1984; and *Huffman et al.*, 1985. The Polar BEAR images have been reported by *Meng and Huffman*, 1987, and by *Huffman et al.*, 1992. The initial work

CHAPTER 16. GLOBAL AURORAL IMAGING

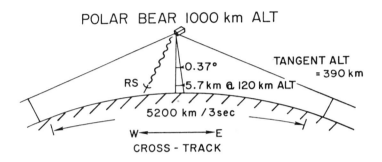

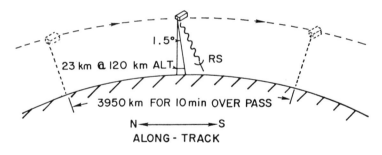

AIRS/POLAR BEAR IMAGE FIELD

Figure 16.3: AIRS/ Polar BEAR imager ground track

16.2. SATELLITE UV IMAGERS

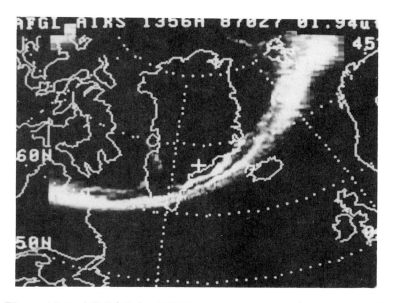

Figure 16.4: AIRS/ Polar BEAR image of aurora (O I 135.6 nm)

on the Polar BEAR data base has been given by *DelGreco et al.*, 1988, and *DelGreco et al.*, 1989. Analysis of auroral oval configurations during quiet conditions has been given by *Meng*, 1989. Image analysis methods used for Polar BEAR auroral images have been described by *Oznovich et al.*, 1992. A typical AIRS image is shown in Figure 16.4.

VIKING The Swedish scientific satellite VIKING was launched on February 22, 1986, into a polar orbit with inclination of $98.8°$, apogee of 13,530 km, perigee of 817 km, orbital period of 262 minutes, and drift of rising node of $0.17°$ per day. There are two extensive publications describing the project and scientific results. A group of 26 short articles under the general heading of "Viking and the aurora" is in the April 1987 *Geophysical Research Letters*, with an introductory paper by *Hultqvist*, 1987. Nine of these papers deal with the UV imager and auroral images. The May 1, 1990, issue of the *Journal of Geophysical Research* contains 27 papers under the heading "Viking investigations of high latitude plasma processes," again introduced by *Hultqvist*, 1990. In these papers, the use of the imager together with the other sensors on the spacecraft is illustrated.

The ultraviolet auroral imager on Viking was provided by a primarily

Canadian team having as principal investigator C. Anger, with scientific collaboration by S. Murphree and others. The primary institution was the University of Calgary. The instrumentation has been described by *Anger, et al.*, 1987a and by *Murphree and Cogger*, 1988. It consists of two cameras with fixed wavelength ranges. One camera isolates the 134–180 nm band, which consists primarily of emission from the N_2 Lyman-Birge-Hopfield bands, and the other covers the 123.5–160 nm band, which is dominated by the atomic oxygen multiplet at 130.4 nm. The 123.5–160 nm band was isolated using a CaF_2 filter, a KBr photocathode deposited on the face of the microchannel plate (MCP), and a reflective coating on the secondary mirror. The 134–180 nm band was isolated with a BaF_2 filter and a CsI photocathode on the MCP. The camera was made by CAL Corp., Ottawa.

The compact camera design uses inverse Cassegrain optics to image the scene on the curved MCP and fiber optics stage, which intensifies the image and provides it to a 288 by 385 pixel CCD. Total field of view is $20°$ by $25°$, the pixel angular field of view is $0.076°$, and the effective pixel size for the detector is 30 $\mu meter$ square. The spatial resolution of the aurora is given as 20–30 km. The cameras have a FOV centered on the orbital plane of the satellite, which is spin-stabilized at three revolutions per minute. Complete auroral images are obtained every 20 seconds, which enables the investigation of many dynamical processes. A method to sum the images over an exposure time per image pixel of one second is utilized before data transmission. The charge accumulating from the scene is synchronized with the image motion due to the satellite spin, which is possible because the CCD columns were held in alignment with the satellite equatorial plane to less than $\pm 0.2°$. The compact design of these cameras is well-suited to use on high altitude satellites, and similar sensors will undoubtedly be flown in the future. One announced use is on Sweden's Freja satellite.

Initial scientific results have been given by *Anger et al.*, 1987b, which also introduces several following papers. There is an extensive data base of these images covering the time period from approximately March to October, 1986. These are being used for further analysis, together with data from the other sensors on the satellite. Since there was no on-board data storage, the images are limited to polar regions in the line of sight of arctic receiving stations. Thus, no images could be acquired of the equatorial ionospheric airglow belts, although the sensor would have been suitable for this use.

EXOS-D (AKEBONO) The Japanese satellite EXOS-D, also called AKEBONO, has as its goal a study of auroral particle acceleration processes

16.2. SATELLITE UV IMAGERS

with measurements from eight sensors, including the ATV, or Visible and UV Auroral Television, sensor. The satellite was launched on February 21, 1989 (UT) into a polar orbit with apogee of 10,500 km, perigee of 274 km, inclination of 75°, and period of 212 minutes. The project manager is K. Tsuruda of the Institute of Space and Astronautical Sciences and the principal investigator for the ATV sensor is T. Oguti of Nagoya University, with scientific collaboration by E. Kaneda of the University of Tokyo and others.

The principal characteristics of the ATV sensor and initial scientific results are described by *Oguti et al.*, 1990. The sensor has two cameras, obtaining images in the visible region in a broad band centered at the oxygen green line (557.7 nm) and in the UV in a band from 115 to 139 nm. The principal contributor to the UV images will therefore be oxygen atom emission at 130.4 nm and 135.6 nm, nitrogen LBH bands, and hydrogen Lyman alpha emission at 121.6 nm. Nonauroral emission will include the intense hydrogen Lyman alpha geocorona. Still images are obtained using a despun mirror on the spin-stabilized satellite. The imager uses image intensifiers and CCD detectors, and the UV imager has a KBr photocathode. The time resolution for a single image can be eight seconds, which is less than previous imagers. Scientific results are beginning to be reported for this satellite at meetings.

FUTURE IMAGERS

The previous discussion of imagers that have been flown establishes that UV imagery is an important tool for auroral studies and eventually for other applications. Although the field is not old, some divergence in the use of imagers and available technology is beginning to become evident.

First, there appear to be two principal divisions of imager usages, which result in different emphases and different technological approaches. It should be pointed out that these are the principal uses only. Suggested names and descriptions of these two different approaches to imagery are as follows:

AURORAL OVAL IMAGERS Auroral oval (AO) imagers are used in programs that seek to study solar-terrestrial relationships involving the magnetosphere and the aurora. In this case, the principal requirement is to obtain the spatial location and image of the aurora so that it can be correlated with other sensors on board the satellite and elsewhere. For these uses, images of the total oval and morphological considerations are more important than the spectral distribution of the emission.

The apogee is usually at 10,000 km or higher, so that important parts of the magnetosphere can be sampled with in-situ sensors. A successive series of snapshots at high temporal and spatial resolution is the primary need. Sensor plans for future VIKING type satellites, the POLAR spacecraft of the International Solar Terrestrial Program (ISTP), and the Soviet INTERBALL program are understood to include UV imagers of this type. A number of others have been proposed, and Section 16.4 will discuss some of the research uses of this type of imager.

AURORAL IONOSPHERIC IMAGERS Auroral ionospheric (AI) imagers are used to understand in more detail the relationships between the aurora, the global ionosphere, and the atmosphere. The spectra of the auroral and airglow emission may be more important than total oval coverage. One goal is to be able to obtain the energy and flux of the precipitating particles from multispectral imagery. Altitudes of from about 800 km up to several thousand km will be used, rather than the higher altitudes of AO imagers. For AI sensors, the imaging capability is combined with the ability to measure airglow emissions with sufficient accuracy to be used as real time inputs in space weather predictive codes. With the same sensors, imaging of the aurora and the ionospheric equatorial airglow belts enable real time definitions of fields and determinations of the current boundaries of auroral, midlatitude, and equatorial ionospheric conditions. These imagers will also obtain or use limb and nadir airglow measurements of spectra for improved composition and density of the local ionosphere. Future planned sensors for the DMSP satellite are of this type. These are discussed further in Chapter 17 as global space weather systems.

A second trend is in the use of array detectors rather than single channel detectors such as photomultipliers. As discussed briefly in Chapter 4, there are many types of array detectors becoming available. They must be tested in experimental space programs, however, before they are considered for long term applications. Factors such as reliability, dynamic range, calibration, complexity, and applications must be considered in the use of sensors in operational programs. Long term costs for data reduction and archiving must be carefully considered for array sensors, which tend to produce a high volume of data. The final applications needs of the users must be kept in mind, rather than research needs. At the present time, the photomultiplier is the detector of choice for operational UV imagers, if at all possible.

16.3. COMPARISON WITH OTHER METHODS

From what we now know, the FUV region (about 200 to 100 nm) will be the primary UV region of interest for auroral imaging. The most useful ultraviolet wavelengths for imagers flown to date have been in the 120 to about 180 nm range. At longer wavelengths, the atmospheric Rayleigh scattering in the daytime is a huge background contribution that prevents good imaging of the aurora. The EUV, from about 100 to 10 nm as defined in this book, is sometimes considered as well. The aurora and equatorial airglow from P78-1 data was analyzed for this purpose by *Meng and Chakrabarti*, 1985. They find that the EUV wavelengths of OII 83.4 nm, OI 98.9 nm, and N_2 100.9 nm are particularly suitable. However, these emissions are weaker than the FUV emissions, and they offer no known advantages compared with the FUV wavelengths. They also require windowless detectors, which is at least a slight disadvantage in an operational sensor compared with the windowed detectors available in the FUV. For these reasons, most imagers will probably continue to use the FUV.

16.3 Comparison with other methods

Space-based methods available for global auroral sensing that do not involve the UV are discussed and compared to UV imaging in this section. These methods may be complementary or competing sensors for use in a given space project. Note that the topside sounder is discussed in Chapter 17, since it is primarily an ionospheric sensor.

A brief description of the major auroral sensors includes:

VISIBLE IMAGERS AND PHOTOMETERS The imagers flown on satellites for auroral imaging have been described earlier in this chapter. The large amount of data available from the Defense Meterological Satellite Program (DMSP) using the visible region cloud imagers at night has been mentioned as well. A recent example involving visible wavelengths is the use of 630 nm measurements from the Atmospheric Explorer program, *Solomon et al.*, 1988.

X-RAY IMAGERS Satellite imaging of the bremsstrahlung, or radiation emitted by precipitating high energy electrons in the auroral zones, has been obtained from several satellites. The energy range of the observed x-rays is in the 1 to 40 keV range, or about 10 to 0.1 Å, although the initial space measurements were made at higher energies, *Imhof et al.*, 1974. Representative measurements from DMSP satellites are described by *Mizera et al.*, 1978. The technique is also discussed by *Walt et al.*, 1979; *Rust and Burstein*, 1981; and *Calvert et al.*, 1985, which describes a sensor flown on the SEEP satellite.

IN-SITU SENSORS The previously mentioned satellite methods are passive remote sensing approaches. Given the current understanding of the aurora, it is possible to use measurements of atmospheric properties at the satellite location, together with field maps and geophysical indices, to provide information about the aurora with the aid of models. These methods are called collectively *in-situ* methods. These measurements include the particle energy and flux and the local electric and magnetic fields. Possibly the best developed correlation of this type at present is the linear relationship between particle measurement of the average equatorial auroral boundaries and the magnetic index K_p (*Hardy et al.*, 1985, 1989). This method of auroral oval location is now used on DMSP satellites. Satellite electric field measurements have also been correlated with visible region auroral locations by *Swift and Gurnett*, 1973.

Visual, x-ray, and UV methods have an advantage over in-situ methods for auroral imaging. The in-situ methods, which obtain measurements along the satellite orbit, must rely on models of fields and statistical ovals to obtain somewhat indirect images of the aurora. On the other hand, by using the electromagnetic radiation emitted by the aurora, the total global image that can be seen from the satellite is possible. Thus, the in-situ sensor gives a 1-D measurement with a modeled 2-D prediction. The imager gives a 2-D map of the emission region directly.

Results from the imagers flown to date are that every auroral image is different, making the use of predictive models incomplete at best. It is difficult to believe that the in-situ approach could ever give an image of a highly disturbed substorm or a theta aurora. In these and other cases, the shape departs drastically from the statistically determined auroral oval.

In-situ sensors are, however, small and well understood. They have been flown a number of times with an extensive data base. In addition, they may be a more sensitive locator of auroral boundaries than the optical methods. The flight of particle sensors together with UV imagers is a good combination, as the data complement each other.

UV imaging is inherently more useful than visible imagery because it can be used both day and night. Visible auroral data is also severely degraded during periods of bright moonlight, which further limits its use due to an

16.3. COMPARISON WITH OTHER METHODS

increased background from surface scattering. As should be apparent from our discussion of the fundamental properties of the UV in the atmosphere, the far (FUV) and extreme (EUV) ultraviolet at wavelengths shorter than about 200 nm, originate at and above ionospheric and auroral altitudes when viewed from space. Thus, the lower part of the atmosphere acts as an absorbing blanket that prevents interference from scattering by clouds or the surface.

X-ray imaging has been compared to UV imaging for application in the auroral ionosphere, *Robinson and Vondrak*, 1987, with the recommendation that the x-ray methods were superior at that time. The x-ray approach, however, only senses radiation from the more energetic electrons, while not being sensitive to 1 to 10 keV electrons and protons that make up the primary energy input to the auroral region. Both x-ray and UV methods will undoubtedly be improved, but at the moment, the highest spatial resolution is with UV methods, since refractive imagery can be used. It may also be difficult to obtain auroral imagery at x-ray wavelengths from the altitudes of several Earth radii needed for some auroral oval (AO) sensors. X-ray imagers do not observe outside the auroral region, which is a further limitation compared to UV approaches, which can observe the equatorial ionospheric airglow belts and airglow intensities for ionospheric electron density modeling.

X-ray imaging does offer possibly the only way to remotely sense the aurora at altitudes lower than 100 km. Again, for research applications, flight of UV and x-ray imagers together would be valuable.

Ultraviolet methods thus have significant advantages for global imaging. The auroral and ionospheric imagery in particular can be used directly for many purposes, beginning with a real time separation of the Earth ionosphere into auroral, mid-latitude, and equatorial regions. The UV offers the potential, also, that airglow and fluorescence measurements from the same sensor, together with models, may allow remote sensing of electron density profiles, atmospheric composition, and atmospheric density, in a global space weather system, as discussed in the next chapter.

A disadvantage or limitation of visible, UV, and x-ray methods is that they observe all of the emission along the line of sight and thus they do not allow the altitude of the emission to be identified. This would be especially useful for separating E and F region emissions. The atmospheric constituents give similar emission whether excited as an airglow, as an aurora, or as an ionospheric irregularity. Possibly multiple imagery with different wavelengths or improved spectral resolution will allow altitude separations to be made. In addition, limb measurements may be coupled with nadir views to obtain some altitude information.

At this stage in the development of space based imagers for the aurora and the ionosphere, it is important wherever possible to fly several types of sensors on research satellites, so that the understanding of the atmosphere can be improved and the characteristics of the sensors can be evaluated.

16.4 Solar-terrestrial physics applications

Ultraviolet imagers are likely to become a standard sensor flown by all space research programs that seek a broad understanding of the complicated pattern of particles, fields, and currents involved in the general research area called solar-terrestrial physics. This research area includes the solar wind, the magnetosphere, the ionosphere, and particle radiation belts, as well as the aurora.

Imagers are frequently used in association with other sensors in coordinated campaigns. Examples of the instrumentation used and representative references are as follows: all sky cameras, *Mende and Eather*, 1976; photometers and spectrometers, *Vallance Jones*, 1974; satellite magnetometers for Birkeland currents, *Potemra*, 1983, *Bythrow et al.*, 1984, 1986; incoherent scatter radar for auroral ionospheric correlations, *Robinson et al.*, 1989, *Oliver et al.*, 1984; and instrumentation carried by aircraft under the auroral oval, *Whalen et al.*, 1971.

Magnetospheric imaging in the UV is a future possibility that should be of great importance in solar-terrestrial research. The magnetosphere is a key part of the sun-Earth system, as was mentioned in Chapter 6. An introduction to the magnetosphere is given by *Roederer*, 1974. The value of auroral imaging in the UV is demonstrated by recent uses in connection with plasma motions in the magnetosphere, *Hill and Dessler*, 1991. Use of the UV scattering by singly ionized atomic oxygen at 83.4 nm has been suggested many times, as discussed by *Swift et al.*, 1989; *Chiu et al.*, 1990; *Meier*, 1990; and references in these papers.

16.5 References

Akasofu, S.-I., Recent progress in studies of DMSP auroral photographs, *Space Science Rev.*, 19, 169–215, 1976.

16.5. REFERENCES

Anger, C. D., T. Fancott, J. McNally, and H. S. Kerr, ISIS- II scanning auroral photometer, *Applied Opt., 12*, 1753–1766, 1973.

Anger, C. D., S. K. Babey, A. Lyle Broadfoot, R. G. Brown, L. L. Cogger, R. Gattinger, J. W. Haslett, R. A. King, D. J. McEwen, J. S. Murphree, E. H. Richardson, B. R. Sandel, K. Smith, and A. Vallance Jones, An ultraviolet auroral imager for the Viking spacecraft, *Geophys. Res. Lett., 14*, 387–390, 1987a.

Anger, C. D., J. S. Murphree, A. Vallance Jones, R. A. King, A. L. Broadfoot, L. L. Cogger, F. Creutzberg, R. L. Gattinger, G. Gustafsson, F. R. Harris, J. W. Haslett, E. J. Lewellyn, J. C. McConnell, D. J. McEwen, E. H. Richardson, G. Rostoker, B. R. Sandel, G. G. Shepherd, D. Venkatesan, D. D. Wallis, and G. Witt, Scientific results from the VIKING ultraviolet imager: An introduction, *Geophys. Res. Lett., 14*, 383–386, 1987b.

Bythrow, P. F., T. A. Potemra, W. B. Hanson, L. J. Zanetti, C.-I. Meng, R. E. Huffman, F. J. Rich, and D. A. Hardy, Earthward directed high-density Birkeland currents observed by HILAT, *J. Geophys. Res., 89*, 9114–9118, 1984.

Bythrow, P. F., M. A. Doyle, T. A. Potemra, L. J. Zanetti, R. E. Huffman, C.-I. Meng, D. A. Hardy, F. J. Rich, and R. A. Heelis, Multiple auroral arcs and Birkeland currents: Evidence for plasma sheet boundary waves, *Geophys. Res. Lett., 13*, 805–808, 1986.

Calvert, W., H. D. Voss, and T. C. Sanders, A satellite imager for atmospheric x-rays, *IEEE Trans. Nuclear Sci., NS-32, No. 1*, 1985.

Carruthers, G. R. and T. Page, Apollo 16 far ultraviolet camera spectrograph: Earth observations, *Science, 177*, 788–791, 1972.

Carruthers, G. R. and T. Page, Apollo 16 far ultraviolet imagery of the polar auroras, tropical airglow belts, and general airglow, *J. Geophys. Res., 81*, 483–496, 1976a.

Carruthers, G. R. and T. Page, Apollo 16 far ultraviolet spectra of the terrestrial airglow, *J. Geophys. Res., 81*, 1683, 1976b.

Carruthers, G. R., T. Page, and R. R. Meier, Apollo 16 Lyman alpha imagery of the hydrogen geocorona, *J. Geophys. Res., 81*, 1664–1672, 1976.

Chiu, Y. T., R. M. Robinson, H. L. Collin, S. Chakrabarti, and G. R. Gladstone, Magnetospheric and exospheric imaging in the extreme ultraviolet, *Geophys. Res. Lett., 17*, 267–270, 1990.

Chubb, T. A. and G. T. Hicks, Observations of the aurora in the far ultraviolet from Ogo-4, *J. Geophys. Res., 75*, 1290–1311, 1970.

DelGreco, F. P., R. E. Huffman, J. C. Larrabee, R. W. Eastes, F. J. LeBlanc, and C. I. Meng, Organizing and utilizing the imagery and spectral data from Polar BEAR, *Ultraviolet Technology II*, R. E. Huffman, editor, *Proc. SPIE 932*, 30–35, 1988.

DelGreco, F. P., R. W. Eastes, and R. E. Huffman, UV ionospheric remote sensing with the Polar BEAR satellite, *Ultraviolet Technology III*, R. E. Huffman, editor, *Proc. SPIE 1158* 46–50, 1989.

Frank, L. A. and J. D. Craven, Imaging results from Dynamics Explorer I, *Rev. Geophysics, 26*, 249–283, 1988.

Frank, L. A., J. D. Craven, K. L. Ackerson, M. R. English, R. H. Eather, and R. L. Carovillano, Global auroral imaging instrumentation for the Dynamics Explorer mission, *Space Sci. Instrum., 5*, 369, 1981.

Frank, L. A., J. D. Craven, J. L. Burch, and J. D. Winningham, Polar views of the earth's aurora with Dynamics Explorer, *Geophys. Res. Lett., 9*, 1001, 1982.

Frank, L. A., J. D. Craven, and R. L. Rairden, Images of the Earth's aurora and geocoronona from the Dynamics Explorer mission, *Adv. Space Res., 5*, 53, 1985.

Fremouw, E. J. and L. A. Wittwer, The HILAT satellite program: Introduction and objectives, *Johns Hopkins/APL Tech. Digest, 5*, 98–103, 1984.

Fremouw, E. J., C. L. Rino, J. F. Vickery, D. A. Hardy, F. J. Rich, R. E. Huffman, C.-I. Meng, K. A. Potocki, T. A. Potemra, W. B. Hanson, and R. A. Heelis, The HILAT program, *EOS, Trans. AGU, 64*, 163, 1983.

Fremouw, E. J., H. C. Carlson, T. A. Potemra, P. F. Bythrow, C. L. Rino, J. F. Vickery, R. L. Livingston, R. E. Huffman, C.-I. Meng, D. A.

16.5. REFERENCES

Hardy, F. J. Rich, R. A. Heelis, W. B. Hanson, and L. A. Wittwer, The HILAT satellite mission, *Radio Sci., 20*, 416–424, 1985.

Gerard, J.-C. and C. A. Barth, OGO-4 observations of the ultraviolet auroral spectrum, *Planet. Spa. Science, 24*, 1059–1063, 1976.

Grant, D. G., Guest editor, The Polar BEAR mission, *Johns Hopkins APL Technical Digest, 8*, 293–339, 1987.

Hardy, D. A., M. S. Gussenhoven, and E. Holeman, A statistical model of the auroral electron precipitation, *J. Geophys. Res., 90*, 4229, 1985.

Hardy, D. A., M. S. Gussenhoven, and D. Bratigan, A statistical model of auroral ion precipitation, *J. Geophys. Res., 94*, 370–392, 1989.

Hill, T. W. and A. J. Dessler, Plasma motions in planetary magnetospheres, *Science, 252*, 410–415, 1991.

Hirao, K. and T. Itoh, Scientific satellite KYOKKO (EXOS-A), *Solar Terr. Env. Res. in Japan, 2*, 148–152, 1978.

Hoffman, R. A., The magnetosphere, ionosphere, and atmosphere as a system: Dynamics Explorer 5 years later, *Rev. Geophysics, 26*, 209–214, 1988.

Huffman, R. E., and C. I. Meng, Ultraviolet imaging of sunlit aurora from HILAT, *Johns Hopkins APL Tech. Digest, 5*, 138–142, 1984.

Huffman, R. E., F. J. LeBlanc, J. C. Larrabee, and D. E. Paulsen, Satellite vacuum ultraviolet airglow and auroral observations, *J. Geophys. Res., 85*, 2201–2215, 1980.

Huffman, R. E., F. J. LeBlanc, D. E. Paulsen, and J. C. Larrabee, Ultraviolet horizon sensing from space, *Shuttle Pointing of Electro-optical Experiments*, W. Jerkovsky, editor, *Proc. SPIE, 265*, 290–294, 1981a.

Huffman, R. E., D. E. Paulsen, F. J. LeBlanc, and J. C. Larrabee, Ionospheric and auroral measurements from space using vacuum ultraviolet emission, in *Ionospheric Effects Symposium 1981: Effect of the Ionosphere on Radiowave Systems*, J. M. Goodman, F. D. Clarke, and J. Aarons, editors, Available from NTIS, 364–371, 1981b.

Huffman, R. E., J. C. Larrabee, F. J. LeBlanc, and C. I. Meng, Ultraviolet remote sensing of the aurora and ionosphere for C3I systems use, *Radio Sci.*, *20*, 425–430, 1985.

Huffman, R. E., J. C. Larrabee, F. P. DelGreco, R. W. Eastes, and C. I. Meng, Ultraviolet imaging and spectra of aurora and airglow from the Polar BEAR satellite, *J. Geophys. Res.*, in preparation, 1992.

Hultqvist, B., The Viking project, *Geophysical Research Lett.*, *14*, 379–382, 1987, and following 26 articles to p. 478 in the section entitled "Viking and the aurora."

Hultqvist, B., The Swedish satellite project Viking, *J. Geophys. Res.*, *95*, 5749–5752, 1990, and following 26 papers to p. 6132 in the section entitled "Viking investigations of high latitude plasma processes."

Imhof, W. L., G. H. Nakano, R. G. Johnson, and J. B. Reagan, Satellite observations of bremsstrahlung from widespread energetic electron precipitation events, *J. Geophys. Res.*, *79*, 565–574, 1974.

Joki, E. G. and J. E. Evans, Satellite measurements of auroral ultraviolet and 3914 Å radiation, *J. Geophys. Res.*, *74*, 4677–4686, 1969.

Kaneda, E., Auroral TV observation by KYOKKO, in *Proceedings of the International Workshop on Selected Topics of Magnetospheric Physics*, p. 15, Japanese IMS Committee, Tokyo, 1979.

Kaneda, E., M. Takagi, and N. Niwa, Vacuum ultraviolet television camera, *Proc. 12th Intl. Symp. Space Tech. and Sci.*, pp. 233–238, Agne, Tokyo, 1977.

Kaneda, E., T. Mukai, and K. Hirao, Synoptic features of auroral system and corresponding electron precipitation observed by KYOKKO, in *Physics of Auroral Arc Formation*, Geophys. Monogr. Ser., Vol. 25, edited by S.-I. Akasofu and J. R. Kan, pp. 24–30, American Geophysical Union, 1981.

Lui, A. T. Y., C. D. Anger, D. Venkatessan, W. Sawchuk, and S.-I. Akasofu, The topology of the auroral oval as seen by the ISIS-2 scanning auroral photometer, *J. Geophys. Res.*, *80*, 1795–1804, 1975.

Meier, R. R., The scattering rate of solar 834 Å radiation by magnetospheric O^+ and O^{++}, *Geophys. Res. Lett.*, *17*, 1613–1616, 1990.

Mende, S. B. and R. H. Eather, Monochromatic all-sky observations and auroral precipitation patterns, J. Geophys. Res., 81, 3771–3780, 1976.

Meng, C. I., Auroral oval configuration during the quiet condition, in *Electromagnetic Coupling in the Polar Clefts and Caps*, P. E. Sandholt and A. Egeland, editors, 61–85, Kluwer Academic Publisher, 1989.

Meng, C. I. and S. Chakrabarti, Extreme ultraviolet emissions for monitoring auroras in dark and daylight hemispheres, *J. Geophys. Res.*, 90, 4261–4268, 1985.

Meng, C. I. and R. E. Huffman, Ultraviolet imaging from space of the aurora under full sunlight, *Geophys. Res. Lett.*, 11, 315–318, 1984.

Meng, C. I. and R. E. Huffman, Preliminary observations from the auroral and ionospheric remote sensing imager, *Johns Hopkins APL Tech. Digest*, 8, 303–307, 1987.

Meng, C. I., R. E. Babcock, and R. E. Huffman, Ultraviolet imaging for auroral zone remote sensing, *AIAA 21st Aerospace Sciences Meeting*, Paper AIAA-83-0019, 1983.

Mizera, P. F., J. G. Luhmann, W. A. Kolasinski, and J. B. Blake, Correlate observations of auroral arc, electrons, and x-rays from a DMSP satellite, *J. Geophys. Res.*, 83, 5573, 1978.

Murphree, J. S. and L. L. Cogger, Application of CCD detectors to UV imaging from a spinning satellite, *Ultraviolet Technology II*, R. E. Huffman, editor, *Proc. SPIE 932*, 42–49, 1988.

Oguti, T., E. Kaneda, M. Ejiri, S. Sasaki, A. Kadokura, T. Yamamoto, K. Hayashi, R. Fujii, and K. Makita, Studies of the aurora dynamics by aurora-TV on the Akebono (EXOS-D) satellite, *J. Geomag. Geoelectr.*, 42, 555–564, 1990.

Oliver, W. L., J. C. Foster, J. M. Holt, G. B. Loriot, V. B. Wickwar, J. D. Kelly, O. de la Beaujardiere, P. F. Bythrow, C. I. Meng, F. J. Rich, and R. E. Huffman, Initial Millstone Hill, Sondrestrom, and HILAT observations of thermospheric temperatures and frictional heating, *Geophysical Res. Lett.*, 1, 911–914, 1984.

Oznovich, I, A. Ravitz, M. Tur, I. Glazer, R. W. Eastes, R. E. Huffman, and A. F. Quesada, Far ultraviolet remote sensing of ionospheric emissions by Polar BEAR, paper in preparation, 1992.

Paresce, F., S. Chakrabarti, R. Kimble, and S. Bowyer, The EUV spectrum of day and nightside aurorae: 800–1400 Å, *J. Geophys. Res., 88*, 4905, 1983.

Potemra, T. A., Magnetospheric currents, *Johns Hopkins APL Tech. Digest, 4*, 276–284, 1983.

Potemra, T. A., Guest editor, The HILAT satellite, *Johns Hopkins APL Technical Digest, 5*, 96–158, 1984.

Robinson, R. M. and R. R. Vondrak, Ultraviolet and x-ray remote sensing of high-latitude ionization from space-based platforms, in *Ionization Effects Symposium 1987: The Effect of the Ionosphere on Communications, Navigation, and Surveillance Systems*, J. M. Goodman, J. A. Klobucher, R. G. Joiner, and H. Soicher, editors, Available from NTIS, 695–704, 1987.

Robinson, R. M., R. R. Vondrak, J. D. Craven, L. A. Frank, and K. Miller, A comparison of ionospheric conductances and auroral luminosities observed simultaneously with the Chatanika radar and the DE-I auroral imagers, *J. Geophys. Res., 94*, 5382–5396, 1989.

Roederer, J. G., The earth's magnetosphere, *Science, 183*, 37–46, 1974.

Rogers, E. H., D. F. Nelson, and R. C. Savage, Auroral photography from a satellite, *Science, 183*, 951–952, 1974.

Rust, D. M. and P. Burstein, Application of x-ray imaging techniques to auroral monitoring, in *Ionization Effects Symposium 1981: Effect of the Ionosphere on Radiowave Systems*, J. M. Goodman, F. D. Clarke, and J. Aarons, editors, available through NTIS, 354–363, 1981.

Schenkel, F. W. and B. S. Ogorzalek, The HILAT vacuum ultraviolet auroral imager, *Johns Hopkins APL Tech. Digest, 5*, 131–137, 1984.

Schenkel, F. W. and B. S. Ogorzalek, Auroral images from space: Imagery, spectroscopy, and photometry, *Johns Hopkins APL Tech. Digest, 8*, 308–317, 1987.

Schenkel, F. W., B. S. Ogorzalek, J. C. Larrabee, F. J. LeBlanc, and R. E. Huffman, Ultraviolet daytime auroral and ionospheric imaging from space, *Appl. Optics, 24*, 3395–3405, 1985.

Schenkel, F. W., B. S. Ogorzalek, R. R. Gardner, R. A. Hutchins, R. E. Huffman, and J. C. Larrabee, Simultaneous multispectral narrow band auroral imagery from space (1150 Å to 6300 Å), *Ultraviolet Technology*, R. E. Huffman, editor, *Proc. SPIE 687*, 90–103, 1986.

Solomon, S. C., P. B. Hays, and V. J. Abreu, The auroral 6300 Å emission: Observations and modeling, *J. Geophys. Res., 93*, 9867–9882, 1988.

Swift, D. W. and D. A. Gurnett, Direct comparison between satellite electric field measurements and the visual aurora, *J. Geophys. Res., 78*, 7306–7339, 1973.

Swift, D. W., R. W. Smith, and S. I. Akasofu, Imaging the earth's magnetosphere, *Planet. Space Sci., 37*, 379–384, 1989.

Vallance Jones, A., *Aurora*, Reidel, 1974.

Walt, M., L. I. Newkirk, and W. E. Francis, Bremsstrahlung produced by precipitation electrons, *J. Geophys. Res., 84*, 967, 1979.

Whalen, J. A., R. A. Wagner, and J. Buchau, Airborne ionospheric and optical measurements of noontime aurora, *J. Atm. Terr. Phys., 33*, 661, 1971.

Chapter 17

Ionospheric Electron Density

The current and predicted ionospheric electron density profile is extremely important in the operation of communications, radar, and navigation systems, which rely on radiowave propagation in the ionosphere. A recent development has been the possibility of obtaining the electron density profile from remote sensing methods based on airglow emission at ultraviolet and visible wavelengths. The principal purpose of this chapter is to describe the current state of this effort.

The preceding three chapters have been concerned with UV remote sensing in the lower atmosphere, in the upper atmosphere, and in the auroral regions. This chapter extends the discussion to the ionized gases of the atmosphere, or the ionosphere, as defined in Chapter 6. Two associated subjects addressed in this chapter are the use of solar UV flux monitoring as an indirect method of ionospheric remote sensing and the UV imaging of ionospheric disturbances at equatorial and at polar latitudes.

Global space weather systems are being developed to monitor ionospheric and atmospheric conditions for the operation of systems using radiowave propagation. Global space weather systems will require UV imaging and radiance measurements from satellites. This chapter concludes with a discussion of the UV methods that will probably be included in these systems, together with related ionospheric modeling and algorithm development. The emergence of global space weather systems for the ionosphere using UV methods has some parallels to the earlier development of stratospheric ozone UV monitoring systems described in Chapter 14.

17.1 Radiowaves and electron densities

Radiowave propagation, so vital to many civilian and military uses, is dependent on the properties of the ionosphere. Radiowaves are propagated through scattering, reflection, and line-of-sight paths. For higher frequencies, the path may be trans-ionospheric, to allow use with high altitude satellites and space probes. Radiowaves, and thus the information that they are carrying, may be distorted by irregularities in the ionospheric composition. It is necessary to understand and measure the properties of the ionosphere in order to use it most effectively.

Our primary concern here is with the *electron density profile*, or the local electron number density as a function of altitude, rather than with the physics of radiowave propagation. The measurement of the electron density profile is key to the improvement of radiowave systems, since it determines critical frequencies and identifies disturbances.

For more information about radiowave propagation and the ionosphere, the recent book by *Davies*, 1989, is recommended. The physics of the ionosphere is discussed by *Rees*, 1989, *Kelley*, 1989, and *Parks*, 1991. The book by Kelley has an introduction to both ground and in-situ measurement techniques in the appendix. A short introduction to the ionosphere is given by *Schunk*, 1983, in a book on solar-terrestrial physics edited by *Carovillano and Forbes*, 1983.

The Handbook of Geophysics and the Space Environment has chapters on ionospheric physics by *Rich and Basu*, 1985, and on ionospheric radio wave propagation by *Basu et al.*, 1985. A excellent introduction to the ionosphere keyed to applications is in the report by *Bilitza*, 1989.

The standard methods for probing the ionosphere are based on the transmission and return of radiowaves. These are important for our purposes because they represent the accepted ground truth measurements to which proposed new remote sensing methods must be compared. Radio scientists around the world routinely measure the height and electron density in the E and F regions of the ionosphere. They pass along this information, in a manner similar to that used for meteorological data, to organizations making radio propagation forecasts.

Ionospheric sounding measurements are made by the following types of sensors:

IONOSPHERIC SOUNDERS The standard ionospheric sounder sta-

17.1. RADIOWAVES AND ELECTRON DENSITIES

tion records the radiowave return as a function of time following transmission of a beam in the zenith direction. The transmitted beam is scanned through a range of frequencies, which are then interpreted to give the height and electron density. Critical frequencies can be readily identified. These are the highest frequencies that can be used at that time and place for many modes of propagation. The typical station scans between 1 and 20 Megahertz, which covers the critical frequencies usually found in the E and F regions. Ionospheric sounder stations are numerous throughout the world, and the abundant data from these stations can be utilized in many ways that are beyond the scope of this book. These are covered in *Davies*, 1989.

INCOHERENT SCATTER RADAR These large facilities have a steerable dish antenna at least 10 meters in diameter. They require a considerable amount of power to operate. For our purposes, their most useful measurement is to map the electron density contours along the scan direction. The data can be used for many other purposes, such as for atmospheric density. The electron density measurements are the best accepted ones for comparison with UV remote sensing techniques, but for this purpose, a coordinated measurement between the satellite and the station must be arranged. There are now incoherent scatter radars in many parts of the globe, but there are still less than ten at the present time (*Bilitza*, 1989). One of the initial arctic stations is described by *Leadabrand et al.*, 1972, and the use of the measurements has been described by *Vondrak*, 1983.

TOPSIDE SOUNDER Radiowave sounders have been used on a number of satellites to remotely sense the topside ionosphere and aurora, as described by *Calvert et al.*, 1964, *Lund et al.*, 1967, and references therein. The book by *Davies*, 1989, has an extensive discussion of topside sounders. The principle of the method is the same as for a ground-based ionospheric sounder. It is an *active* as opposed to a *passive* remote sensor, since it is necessary to transmit and receive the radio wave from a sensor on board the satellite. Use of an active method on a satellite usually requires more weight and power. In addition, the side effects of potential radio frequency interference place severe demands on the shielding and fabrication of the satellite. As a research tool, however, satellite topside sounders have allowed better understanding of the topside of the ionosphere, which cannot be studied with ground-based sounders.

As frequency is scanned on an ionosonde, return signal is obtained resulting from reflection of the transmitted signal from the ionosphere. However, at the *critical frequency*, the return decreases greatly, as the transmitted signal passes through the ionospheric layer rather than being reflected. There are critical frequencies measurable for the E, F_1, and F_2 ionospheric layers in the daytime. The critical frequency is thus the highest one that can be used for reflection, or skip-path, propagation. Transmissions at higher frequencies using the reflection property of the ionosphere are not possible.

These frequencies are related to the maximum or peak electron density of the ionospheric layer through the following equation:

$$n_{max} = 1.24 \times 10^4 f_o^2, \qquad (17.1)$$

where n_{max} is the peak electron density of the layer in units of $electron/cm^3$ and f_o is the critical frequency in Megahertz. The critical frequencies are usually designated by the layer involved, with the F_2 layer frequency, for example, being called f_oF_2. The subscript on the frequency means that the frequency applies to the ordinary trace.

The total electron content (TEC) of the ionosphere, or the total column density of electrons, is of importance in transionospheric propagation. It can be obtained by integrating the electron density profile along the transmission path.

The ionospheric regions are modified Chapman layers as introduced in Chapter 7. There are no separate, clearly defined layers in the ionosphere, of course, but the concept is widely employed. Normally, the peak electron densities increase in the layer sequence E, F_1, and F_2 (see Figure 6.4). The ionosonde data for each layer is most useful at frequencies up to the critical frequency, and the profile above the peak in each layer is not measured by the ionosonde. In the topside case, the returns are for the electron density profile down to the f_oF_2 peak, and the ionosphere below this peak is not measured.

Use is also made of radiowave beacons, or transmitters, on satellites to study propagation conditions. The application of this method to study the scintillations produced in radiowaves during trans-ionospheric propagation by ionospheric irregularities is discussed in Section 17.4.

17.2 Electron densities from UV radiance

The UV airglow and scattered radiation emitted by the atmosphere at ionospheric altitudes is being studied for use in passive remote sensing methods for electron density profiles and other ionospheric characteristics. If optical emission could become the basis for this method of remote sensing, there would be a number of advantages, including relatively straightforward adaption to operational systems in space.

Day and *night* ionospheres are discussed separately, as the sensing methods are different. The day ionosphere discussion is further limited to the *day midlatitude* case, which is defined here to be the latitudes between the equatorward edge of the aurora and the onset of anomalous effects centered on the magnetic equator.

The *polar regions* containing the auroral zone and the polar cap must be treated separately. The E-layer in a diffuse, or continuous, aurora has been modeled, but the chaotic F-region occurring at higher altitudes is unlikely to have a predictive model soon. The *equatorial ionization regions* also require different methods, as the UV emission is from ion-electron recombination.

The ionospheric sensing methods are all somewhat indirect and depend on the use of models. Absorption and occultation methods are not possible. Also, there is no direct emission from an ionized constituent. Thus, the consideration of passive remote sensing by optical means is tied to our understanding of the total global ionospheric system. While it is reasonable to believe that these methods can be made to work, and it is certain that they possess operational advantages over other approaches, their acceptance as a method to obtain electron density profiles, critical frequencies, and total electron content will require a considerable amount of additional research.

DAY MIDLATITUDE IONOSPHERE

Space-based UV remote sensing of the midlatitude ionosphere in the daytime has primarily centered on two approaches which we will call the **ab initio** method and the **83.4 nm** method.

The **ab initio** approach uses the airglow and scattering emission, observed in the nadir or near-nadir direction, as boundary conditions for comprehensive models of the ionospheric regions of the atmosphere. The models are adjusted until the calculated emission matches the measured emission. When this adjustment is complete, the model will provide the electron density profile, total electron content, and other atmospheric properties. The aim is to provide ionospheric predictions to the accuracy necessary with the fewest airglow measurements in a relatively simple operational system.

The *ab initio* approach can be readily combined with UV auroral imaging techniques described in the last chapter. The goals of the AIM imager on the HILAT satellite and the AIRS imager on the Polar BEAR satellite, which were described in the previous chapter, included the measurement of airglow emission for the purpose of developing improved ionospheric remote sensing. The *ab initio* approach can also incorporate data from in-situ sensors of particle precipitation and local electron density, to increase the accuracy of the predicted electron density profile. The use of additional UV and visible airglow bands will lead to greater accuracy, if needed. In addition, the method lends itself to incorporation of solar EUV flux measurements, when they become available.

The *ab initio* approach has grown out of both theoretical and experimental reseach on the ionosphere and the aurora. The electron density profile in the *auroral E layer*, the relatively structureless ionosphere associated with the diffuse aurora, was shown with model calculations to be obtainable using a limited number of UV emissions such as atomic oxygen, O, 135.6 nm; molecular nitrogen, N_2, Lyman-Birge-Hopfield (LBH) bands; and nitrogen ion, N_2^+, 391.4 nm, by *Strickland*, 1981, and *Strickland et al.*, 1983.

The theoretical framework for the use of this method is based on earlier work on atmospheric and ionospheric modeling by *Jasperse*, 1977, 1981, and by *Anderson*, 1973. The method requires a massive amount of data on cross sections, reaction rates, energy levels, and associated quantities, as demonstrated by the report of *Wadzinski and Jasperse*, 1982. Improved laboratory measurements must be incorporated into the *ab initio* approach as they become available. However, over time the most significant quantities can be identified through sensitivity analyses, and measurement needs can be better focused.

The modeling studies were based in part on airglow and auroral measurements from the S3-4 satellite, *Huffman et al.*, 1980, and their application to auroral and ionospheric sensing, *Huffman et al.*, 1981. The VUV Backgrounds experiment on the S3-4 satellite is described more completely in Chapter 4.

The *ab initio* modeling efforts have continued, *Strickland and Daniell*, 1982; *Strickland et al.*, 1984; *Daniell et al.*, 1985; *Decker et al.*, 1986, 1987, 1988, 1990. The last paper uses satellite in-situ plasma measurements only, to indicate their applicability for incorporation with UV airglow measurement into the *ab initio* approach. This general approach is also recently

17.2. ELECTRON DENSITIES FROM UV RADIANCE

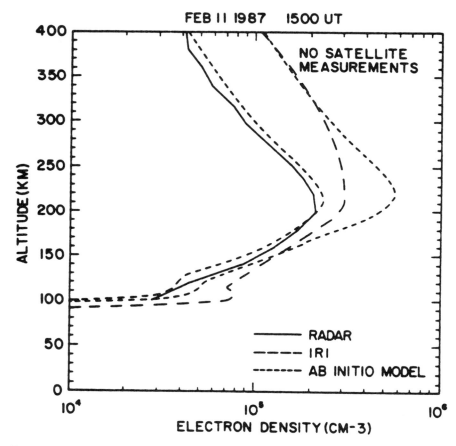

Figure 17.1: Electron density profiles based on the *ab initio* approach with no satellite measurements (*Decker et al.*, 1988)

referred to as the SATMAP method.

Primarily experimental work relating to the development of this method has been reported by *Huffman et al.*, 1984, 1985; *Meng et al.*, 1986; and *DelGreco et al.*, 1989. Comparison with x-ray methods in auroral regions was made by *Robinson and Vondrak*, 1987.

An example from modeling studies of the electron density profile is given in Figures 17.1 and 17.2. In Figure 17.1, no satellite data records are used and the envelope of the model calculations lies higher than the electron

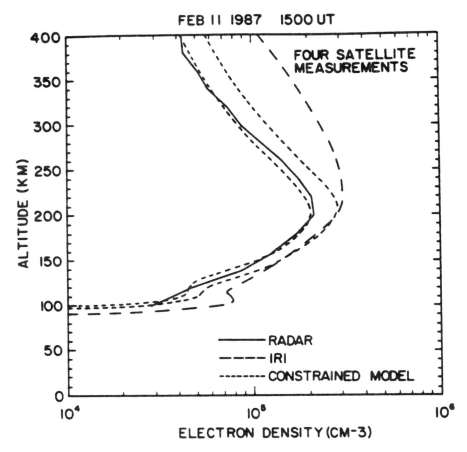

Figure 17.2: Electron density profiles based on the *ab initio* approach with UV airglow and in-situ satellite measurements (*Decker et al.*, 1988)

densities predicted by the International Reference Ionosphere (IRI) and the incoherent scatter radar measurement. When four satellite measurements are included in Figure 17.2, the envelope of the model calculations is in much better agreement. In this case, both UV airglow and in-situ measurements were used. Both figures are for the day midlatitude case.

The *ab initio* approach may be criticized as being too indirect and based too much on modeling. The airglow emissions used do not originate from ions or electrons, but from collision processes of the photoelectrons with

17.2. ELECTRON DENSITIES FROM UV RADIANCE

atmospheric constituents. As the progress on comprehensive atmospheric models proceeds, this objection should be less important. UV airglow serves as an observable that may carry enough information to serve as the primary boundary values in models. In order to use this method, accurate airglow intensity measurements are needed, so adequate plans to keep the instruments in calibration throughout their flight are needed.

Current work on the *ab initio* method includes the definition of the best wavelength bands to use and the development of algorithms to readily incorporate the measurements into predictive models. The earlier plan to use only the 135.6 nm and an LBH band may have been too limited. One current plan is to use five wavelength bands, including atomic oxygen 135.6 and 130.4 nm, two LBH bands, and another emission band such as atomic hydrogen 121.6 nm (Lyman α).

A satellite spectrometer experiment to provide further data for the approach is the AURA (Atmospheric Ultraviolet Remote Analyzer) experiment, now being developed by the Phillips Laboratory for flight in about 1994. It will obtain FUV radiance measurements with two spectrometers in coordination with ground truth electron density measurements by sounders and radars.

Useful operational systems may well include other wavelengths, possibly observed in the limb to avoid solar scatter. The 391.4 nm emission from the molecular nitrogen line has been mentioned. A recent paper on the 630.0 nm atomic oxygen emission as observed in the dayglow limb indicates that other wavelengths in the visible region should be considered, *Solomon and Abreu*, 1989.

The **83.4 nm** method of ionospheric sensing depends on the scattering of radiation from atomic oxygen ions in the F-region of the ionosphere. In this two-step process, solar EUV photons absorbed at altitudes lower than the F-region result in the emission of 83.4 nm emission through direct solar photoionization and through impact with photoelectrons. This emission then illuminates the F-region from below, resulting in resonance scattering by the O^+ in the ionosphere. The electron concentration is equal to the ion concentration, which at these altitudes is largely O^+.

Measurements and modeling of this emission for the purpose of O^+ determination were reported using STP 78-1 satellite data by *Kumar et al.*, 1983. The report by *Meier*, 1984, and the atmospheric model by *Anderson and Meier*, 1985, provide further theoretical understanding needed for this method. A parametric study has been done by *McCoy et al.*, 1985. This

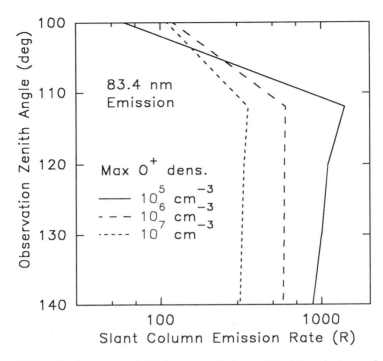

Figure 17.3: Limb scans of 83.4 nm emission related to electron density, based on *Meier*, 1991

study leads to the possibility that the shapes of the limb profiles of 83.4 nm emission could be used rather than the airglow measurements themselves. The data from the STP 78-1 satellite were not sufficiently detailed to examine this possibility adequately.

Experimental measurements to further test this approach should come from the many limb airglow measurements planned on the satellite experiment called RAIDS (Remote Atmospheric and Ionospheric Detection System), *McCoy et al.*, 1986, 1987, and *Paxton and Strickland*, 1988. Although launch dates tend to slide, it is anticipated that measurements from this source will be available before the mid-90s.

The *83.4 nm* approach has been further explored recently by *Meier*, 1991, using improved laboratory measurements. Limb emission scans based on his calculations are shown in Figure 17.3.

17.2. ELECTRON DENSITIES FROM UV RADIANCE

The *83.4 nm* method requires detailed models of the atmosphere. One difficult part of the modeling is the absorption by molecular nitrogen of 83.4 nm radiation from the lower atmosphere as it is transmitted to the F-region to be scattered by O^+. The emission referred to as 83.4 nm is actually a multiplet with three closely spaced lines, and the nitrogen absorption cross section is so structured in this region that the cross sections are very different for the individual members of the multiplet. The overlap of the emission line and the absorption line typically occurs on the steep sides of the lines, where changes due to temperature and collisions should be greatest. This aspect of the problem has been discussed by *Cleary et al.*, 1989, and new cross section measurements have been given by *Morrison and Cleary*, 1989. The recognition of the complexity in this part of the problem demonstrates the necessity of detailed understanding of all aspects of the atmosphere.

At this stage of the development of UV remote sensing methods for the ionosphere, both the **ab initio** and the **83.4 nm** approaches appear feasible. However, comparison with ionosonde and incoherent scatter radar data is virtually nonexistent. It is to be expected that transport processes such as winds and other global circulation effects, which have been left out of the models in any detail thus far, may well determine the better approach.

Until sufficient flight measurements are made in coordination with ionospheric sounders and radars, both approaches must be considered at this time to be unvalidated suggestions. More experimentation, followed by more modeling, is required.

NIGHT MIDLATITUDE AND EQUATORIAL IONOSPHERE

Electron densities become small in the night ionosphere, and ionization sources weak compared with solar photoionization become important, as discussed by *Strobel et al.*, 1974, 1980. The F_1 and F_2 layers found in the daytime merge to a single F-region, and the D and E layers virtually disappear. Near the equator, the night ionosphere is dominated by ionization belts on either side of the magnetic equator. Ultraviolet emission in these equatorial airglow belts, described in Section 9.3, is the basis for UV remote sensing of the critical frequency, f_oF_2, and the height of the F-region maximum, H_M.

The principal emission from the equatorial, or tropical, airglow belts is due to radiative recombination of atomic oxygen ions and electrons in the F-region, leading to many emission features dominated by the atomic oxygen 130.4 and 135.6 nm multiplets. The emission process is thus much simpler than in the daytime case, and the observed emission can be tied with the F-region critical frequency, f_oF_2, through Eq. 17.1.

Several fortuitous near overpasses of ionosondes were compared with OGO-4 satellite data from ionization cells by *Meier and Opal, 1973*. Although there was considerable scatter in the values of f_oF_2 obtained by the two methods, the results were promising, and they still represent one of the few cases of a comparison of UV measurements from space with ionosonde data.

A more comprehensive approach to night ionospheric remote sensing is based on *Chandra et al., 1975*, and *Tinsley and Bittencourt, 1975*. The method is described in detail in these papers. It is based on the measurement of atomic oxygen 135.6 nm and the atomic oxygen 630.0 nm emission. From these measurements, the F-region maximum electron density and its height can be obtained. It is feasible to consider measurement of the 630.0 nm red line in the nadir direction at night, but interference by ground scattering at all times and by lunar scattering from the ground over parts of each month make limb measurements more attractive. A reexamination of the emission of 630.0 nm has been given *Link and Cogger, 1988*.

While the method discussed in the last paragraph is generally accepted as suitable for a remote sensing method, it must be remembered that it has not had an experimental test. As with the daytime methods, coordinated satellite airglow and ground-based sounder and radar measurements are needed. There is also the question about its use in the midlatitude region, where the 135.6 nm airglow intensity is very weak (see Figure 9.4).

An additional suggestion for night ionospheric sensing was made by *Gerard et al., 1977*, who propose use of the ratio of the atomic oxygen emissions at 135.6 and 130.4 nm for F-region height determination. This method should also be examined in future experiments.

Another emission that may prove useful in the future is the *radiative recombination continuum of atomic oxygen*, which emits over a small wavelength region shortward of the ionization threshold at 91.1 nm (see Section 12.2). This emission comes only from radiative recombination and is not influenced by cascading from higher energy levels in the oxygen atom. Use of all of the nighttime emissions in a practical system will become more feasible when larger aperture sensors in space can be considered.

The above brief discussions indicate that there are proposed UV-visible optical methods for ionospheric remote sensing of major global regions. Work continues to find methods for the ionosphere in the aurora and the polar cap,

17.3. UV SOLAR FLUX APPROACHES

as mentioned in various places in this book.

The ionospheric remote sensing methods based on UV airglow and scatter emission are promising, but at this time they lack detailed comparison with ground truth, or accepted, methods for ionospheric monitoring, which are ionosonde and incoherent scatter radar measurements. The acquisition of comprehensive data sets from space is needed, with enough time in orbit available to observe the various disturbance conditions that occur in the ionosphere. The advantages, probably including cost in the long run, of satellite remote sensing remain attractive, and the passive approach makes incorporation into multisensor satellites relatively easy. Current plans for operational sensors are described in Section 17.5.

17.3 UV solar flux approaches

Solar flux measurements have been considered as an input to global circulation models which will yield the electron density profile and other ionospheric characteristics, along with many other atmospheric properties.

Local electron densities have been calculated from solar flux measurements made by sensors on sounding rockets using a simplified ionospheric model, *Heroux et al.*, 1974, 1975. This work indicated the promise of ionospheric sensing using solar flux data, but it was not suited for global use. It did, however, stimulate other remote sensing efforts in the ionosphere.

The key measurements for this approach are the solar EUV and FUV irradiance. The currently available flux values, primarily due to Hinteregger and associates, have been discussed in Chapter 7. The EUV spectrum of interest has been divided into 37 lines and regions, as given most recently by *Torr and Torr*, 1985. At the present state of development of remote sensing models, it is considered necessary to measure these lines individually, rather than to use some wide bandwidth approach. A useful catalog of measurements and models available through the National Space Science Data Center is by *Bilitza*, 1990.

Solar EUV and FUV flux measurements are inputs to large *global circulation models* that have as one of their outputs the electron and ion number densities. A discussion of models of this type is given by *Roble et al.*, 1987. These models also give the neutral composition and are therefore closely related to the atmospheric density applications of Section 15.5. The EUV flux varies greatly with solar activity compared to the FUV flux. It is a frequent comment in papers on ionospheric models that improved and continuing measurement of the solar EUV flux are needed, but unfortunately this type of routine measurement is probably a number of years into the future.

Comprehensive ionospheric and thermal models require knowledge of the reactions and collision processes occurring. Examples of this complexity are given by *Richards and Torr*, 1986, 1988, and references therein. In the earlier work, the F-region peak electron density is found to be sensitive to enhanced vibrational excitation of molecular nitrogen during solar maximum. In the latter work, the need to include ionization by solar photoelectrons as well as by direct absorption of solar EUV is stressed. Work of this type that seeks to define the significant atmospheric processes and that emphasizes the need for their accurate measurement in the laboratory continues to be valuable in shaping the direction of the field.

17.4 Ionospheric irregularities

Equatorial and *polar* irregularity structures are found in the ionosphere. These structures are enhancements or depletions of the local electron density, and they are found imbedded in the ambient ionosphere. The importance of irregularities for our discussion is that they generally have a detrimental effect on the operation of radiowave systems, as discussed by *Carlsen*, 1987, and by *Davies*, 1989 (especially Chapters 8 and 9). Effects on synthetic aperture radars are discussed by *Szuszczewicz et al.*, 1983. Irregularities and disturbances may lead to fade-out, phase and amplitude scintillations, and other noise and distortion of the signal. Thus, methods for identifying and predicting irregularities are of practical importance. It is possible that UV methods can be used for this purpose.

The nontypical ionospheric conditions discussed in this section are known collectively as *irregularities* or sometimes *disturbances*. Many ionospheric irregularities are due to solar activity. Some of these are sudden ionospheric disturbances, polar cap absorption events, and magnetic storms or substorms. Other irregularities are called patches, bubbles, plumes, blobs, depletions, and sun-aligned arcs. They are known or are suspected to be associated with sporadic E-layer formation and spread-F propagation conditions. Some may be caused by meteor showers or cosmic rays. The reader is referred to the general references on the ionosphere given earlier and to the review by *Fejer and Kelley*, 1980, for additional information.

The name of *scintillation* or *beacon* measurements is given to a radiowave

17.4. IONOSPHERIC IRREGULARITIES

method used for the study of irregularities. In this method, the phase and amplitude fluctuations, or scintillations, in a radiowave signal transmitted through the ionosphere are mathematically analyzed into frequencies. These frequencies, displayed as power spectral densities, are used for further theoretical study and analysis.

The Defense Nuclear Agency (DNA) Wideband (1977), HILAT (1983), and Polar BEAR (1986) satellites have all carried radio beacons and analyzed the data in the manner indicated. The DNA applications interest was to study natural ionospheric irregularities as an example of the radiowave propagation difficulties to be expected after atmospheric nuclear detonations.

The Wideband satellite, *Fremouw et al.*, 1978, was largely concerned with equatorial irregularities, while HILAT and Polar BEAR primarily investigated the polar regions. HILAT and Polar BEAR carried UV imagers, as discussed in the last chapter, with one goal the identification of regions causing scintillation.

We are concerned here with natural ionospheric irregularities only. The artificial modification and study of the ionosphere with active experiments such as chemical releases, ionospheric heaters, explosions, or charged particle beam injection methods is outside our scope. Ultraviolet techniques, which in these cases would be mainly from space, have not been applied to these interesting experiments to any extent. The availibility of space imagery may change this situation.

EQUATORIAL IRREGULARITIES These are found associated with the so-called anomolous ionosphere located largely in the equatorial region and near the magnetic equator.

In the equatorial region, bubbles or depletion regions form, as seen graphically in incoherent scatter radar data. Optical imaging using the atomic oxygen 630 nm and other airglow emissions has led to knowledge of the morphology and movement of these structures. These observations are discussed by *Tinsley*, 1982; *Moore and Weber*, 1981; *Weber et al.*, 1983; *Rohrbaugh et al.*, 1989; and references therein.

Ultraviolet spectra of the equatorial ionosphere, obtained primarily at night from space, indicate that the airglow associated with the radiative recombination of atomic oxygen ion is the primary emission. As discussed in Section 17.2, these emissions are closely related to the electron density.

While there is as yet no UV image from space of a depletion region, it is apparent that depletion regions may lead to a local decrease in the UV airglow as seen from space. This effect, if large enough, could be used as the basis for an imaging method. The atomic oxygen emissions at 130.4 nm, 135.6 nm, and possibly other lines in these series could possibly be used, as could the recombination continuum near 91.1 nm.

POLAR IRREGULARITIES In the polar regions, the primary observable in the ultraviolet from space is the auroral oval, as covered in detail in the last chapter. There are several kinds of regions of enhanced ionization known in the polar cap, at latitudes higher than the oval, and also in the auroral zone itself. These regions of enhanced ionization are largely at F-region altitudes, while the auroral emission from particles is largely from E-region altitudes.

The polar ionized structures, their formation and decay, and their movement across the region have been the subject of much research. In the most fruitful investigations, a number of diagnostic techniques are brought to bear on their observation. A comprehensive measurement program of a polar cap arc is reported by *Weber et al.*, 1989, which also gives references to the extensive work in this field.

At the present time, study of polar irregularities using UV imaging from space is just beginning. Use of radio, radar, and Polar BEAR satellite images to characterize auroral E-region plasma structures has been completed, *Basu et al.*, 1991. As can also be seen in Figure 2.5, sun-aligned arcs are in the FUV Polar BEAR images. Polar cap structure has been reported in the EUV, *Chakrabarti*, 1985. The further investigation of these structures with improved UV imagers added to current instrumentation should prove important.

17.5 Global space weather systems

A *global space weather system* will provide current ionospheric and atmospheric conditions, and forecasts of future conditions, based on global measurements and models. This system would combine relevant ground and space measurements and computational codes to allow ionospheric and other atmospheric conditions to be provided to users on a continuing basis. Such systems are in the planning stage at this time, and it is quite likely that they will include ultraviolet sensors in space as a valuable element. Our discussion here will be largely restricted to UV methods used in space and the associated models.

The output products from the space weather system will be tied to needs

17.5. GLOBAL SPACE WEATHER SYSTEMS

for radio communication, radar, radio navigation, and other activities. One output product would probably be the critical frequency f_oF_2 at any geographic location and possibly a forecast of its future value. Other standard data items and indices obtained from the operational software must be developed, distributed, and accepted for use by customers. This acceptance by users is the completion of a technology transition process.

Some idea of the disruption to our normal activities due to space weather ionospheric effects following solar storms, such as in March 1989, can be obtained from *Allen, et al.*, 1989.

SATELLITE UV SENSORS Global space weather system satellites will use UV remote methods for the thermosphere, aurora, and ionosphere as discussed in Chapters 15 and 16 and earlier in this chapter. Airglow and auroral imaging and radiance measurements in both Earth-directed and Earth-limb viewing directions will be included. Solar flux measurements from the same satellite would be a valuable addition.

The space weather system could well provide nonionospheric sensing with the same or similar sensors. In particular, it may be possible to include atmospheric composition, density, and temperature measurements. Also, the use of multispectral auroral imagery to provide the flux and energies of precipitating particles in the auroral zone, discussed in Section 10.3, may be possible. It is expected that in-situ and other presently used sensors will be incorporated.

A number of the above measurements are included in two ultraviolet sensors being developed for operational use by the DMSP office of the Air Force's Space System Division. These are a nadir imaging sensor named SSUSI, or special sensor ultraviolet spectrographic imager, and a limb altitude profile sensor called SSULI, or special sensor ultraviolet limb imager. They are being developed in coordination with algorithms and models for flight in the late 1990s. Although these plans are subject to change, the current plans have the nadir sensor using imaging wavelengths of OI, 135.6 nm; N_2 LBH bands in broad wavelength regions around 140 and 165 nm; H Lyman α, 121.6 nm; and OI, 130.4 nm, in rough priority order. The limb instrument would cover the O^+ emission at 83.4 nm; OI 135.6 nm; and some LBH bands as a minimum.

IONOSPHERIC MODELS A synthesis now taking place that will ultimately be vital to global space weather systems is to enlarge existing ionospheric models to include real time observations both from ground sounder networks and from satellite sensing in the UV and elsewhere. In addition, the physical and chemical processes of the ionosphere will eventually also be included.

An introduction to the worldwide ionospheric data base, current models, and sources of ground and space based measurements, is summarized in a useful report by *Bilitza*, 1989. The International Reference Ionosphere 1990 is described by *Bilitza et al.*, 1990. These two reports serve as sources for references to earlier ionospheric modeling. Another very useful review of the current status of various ionospheric modeling approaches is by *Schunk and Sojka*, 1992.

The models are being enlarged to calculate or use as an input UV and optical measurements. Development of individual techniques such as the *ab initio* method and the *83.4 nm method* in the day midlatitude have been previously covered. Models of this kind will become modules in more comprehensive global ionospheric models.

Examples of the use of optical data such as images and airglow intensities in ionospheric models can be found in the review by *Sojka*, 1989; in the use of Dynamics Explorer UV auroral images for F region dynamics, *Sojka et al.*, 1989; in the testing of the SLIM model against satellite 630 nm airglow measurements, *Anderson et al.*, 1987; in similar tests using 630 nm airglow data from ground stations, *Sahai et al.*, 1990; in the combination of satellite airglow and ground measurements, *Serafimov and Serafimova*, 1992; and in other references found in these papers. The importance of a global perspective in future ionospheric modeling is well recognized, *Szuszczewicz*, 1986.

Computer models and codes directed toward incorporation into the DOD weather system plan to use UV and optical approaches. As an example of this activity, work on ionospheric and magnetospheric specification models can be cited, *Tascione et al.*, 1988, and references therein. A significant feature from our perspective apparently first introduced for the **ICED** (Ionospheric Conductivity and Electron Density) model, as described in the above paper, is the use of real time satellite (DMSP) visible imager data in an ionospheric model. This recognition of the value of auroral imaging in the visible region for ionospheric modeling makes the use of UV imaging from the DMSP more appealing. The aurora can be imaged in the UV both day and night, while good visible imagery is only possible during night conditions with little moonlight.

The value of UV space imagery of the aurora, polar cap, and equatorial ionospheric emission belts has become recognized, and recent models

17.5. GLOBAL SPACE WEATHER SYSTEMS

are planning to utilize more imager data in real time. Theoretical models cannot tell where polar cap, auroral, midlatitude, and equatorial anomaly conditions exist at any given time, and the physics of the problem changes in each of these regimes. One current approach is to use arbitrary separations into low latitude, $0-25°$; midlatitude, $25-50°$; and high latitude, $50-90°$. Ultraviolet imagery allows real time definition of most of the important boundaries separating ionospheric regions, such as between auroral and midlatitude conditions. These key boundaries would define the regions of applicability of different parts of the models, serving as flags in the input information defining which algorithms should be used at any point on the globe.

Note that the use of the imagery data is separate from and independent of the use of airglow and auroral intensities at different spectral features for determination of electron density profiles, atmospheric species concentrations, and associated properties. This additional data can, however, be readily obtained from the imagers, so that a considerable amount of information is obtained from an ultraviolet imager sensor on an operational satellite. The best imagery is from a primarily nadir viewing sensor, but limb radiance measurements provide altitudinal coverage not available from the nadir sensors.

The ICED model has been extended with the development of a model called **GLOBAL ICED**, as described by *Daniell et al.*, 1990. The initial ICED model covered the northern polar region only, while GLOBAL ICED also covers the remainder of the globe. Typically, the most useful of previously developed models are incorporated into more comprehensive models such as GLOBAL ICED. A complete review of the component models will not be attempted here.

Work on the models indicates the value of a semi-empirical approach using measurements from all available space and ground sources. The current name for a comprehensive model is **PRISM**, or Parameterized, Real-time, Ionospheric Specification Model, which is being developed by the Phillips Laboratory, Geophysics Directorate, for the DOD Space Forecasting Center.

Algorithm development is a key part of the codes. The algorithms will convert the measurements, or sensor data records, to geophysical quantities such as electron densities, called geophysical data records. These algorithms must then be incorporated into operational software for the system.

The actual system placed in operation will of course be determined by the perceived needs at the time. The specific measurements and experimental techniques have to be chosen. The goal is to achieve the needed user products with the minimum amount of hardware and difficulty.

17.6 Concluding remarks

The discussion of ultraviolet methods for determining ionospheric electron density given in this chapter concludes this overview of the ultraviolet and its atmospheric sensing applications. Our discussion began with the fundamentals of ultraviolet sensors and radiometry, continued with a description of the Earth's atmospheric environment from the UV perspective, and concluded with applications and potential applications. It is hoped that the inclusion of both atmospheric and ultraviolet topics in a single book will be found useful.

In many cases, the discussion could be greatly enlarged, but this detail is not possible given space and time limitations. Numerous references to the literature have been included, which should enable specialists to conduct a more thorough study.

17.7 References

Allen, J., L. Frank, H. Sauer, and P. Reiff, Effects of the March 1989 solar activity, *EOS, Transactions, American Geophysical Union, 70*, 1479–1488, November 14, 1989.

Anderson, D. E., Jr., and R. R. Meier, The OII 834 Å dayglow: A general model for excitation rate and intensity calculations, *Planet. Space Sci., 33*, 1179, 1985.

Anderson, D. N., A theoretical study of the ionospheric F region equatorial anomaly, 1, Theory, *Planet. Space Sci., 21*, 409, 1973.

Anderson, D. N., M. Mendillo, and B. Herniter, A semi-empirical low latitude ionospheric model, *Radio Sci., 22*, 292, 1987.

Basu, S., J. Buchau, F. J. Rich, E. J. Weber, E. C. Field, J. L. Heckscher, P. A. Kossey, E. A. Lewis, B. S. Dandekar, L. F. McNamera, E. W. Cliver, G. H. Millman, J. Aarons, J. A. Klobuchar, and M. F. Mendillo, Ionospheric Radio Wave Propagation, Chapter 10 in *Handbook of Geophysics and the Space Environment*, A. S. Jursa, Scientific Editor, Air Force Geophysics Laboratory, NTIS Document Accession Number ADA 167000, 1985.

17.7. REFERENCES

Basu, Sunanda, S. Basu, R. W. Eastes, R. E. Huffman, R. E. Daniell, P. K. Chaturvedi, C. E. Valladares, and R. C. Livingston, Remote sensing of auroral E-region plasma structures by radio, radar, and UV techniques at solar minimum, paper submitted to *J. Geophys. Res.* for publication, 1991.

Bilitza, D., *The Worldwide Ionospheric Data Base*, NSSDC/ WDC A-R&S, 89-03, National Space Science Data Center, NASA Goddard SFC, 1989.

Bilitza, D., *Solar-Terrestrial Models and Application Software*, NSSDC/ WDC A-R&S, 90-19, National Space Science Data Center, NASA Goddard SFC, 1990.

Bilitza, D., with contributions by K. Rawer, L. Bossy, I. Kutiev, K.-I. Oyama, R. Leitinger, and E. Kazimirovsky, *International Reference Ionosphere 1990*, NSSDC/ WDC A-R&S 90-22, National Space Science Data Center, NASA Goddard SFC, 1990.

Calvert, W., R. W. Knecht, and T. E. VanZandt, Ionosphere Explorer I satellite: First observations from the fixed frequency topside sounder, *Science, 146*, 391–395, 1964.

Carlson, H. C., Jr., A C^3I view of auroral and polar cap ionospheric phenomenology, in *Ionospheric Effects Symposium 1987: Effect of the Ionosphere on Communication, Navigation, and Surveillance Systems*, Editors: J. M. Goodman, J. A. Klobucher, R. G. Joiner, and H. Soicher, Available from NTIS, 617–624, 1987.

Carovillano, R. L. and J. M. Forbes, editors, *Solar-Terrestrial Physics*, Reidel, 1983.

Chakrabarti, S., EUV (300–900 Å) spectrum of polar cap and cusp emission near local noon, *J. Geophys. Res., 90*, 4421–4426, 1985.

Chandra, S., E. I. Reed, R. R. Meier, C. B. Opal, and G. T. Hicks, Remote sensing of the ionospheric F layer by use of OI 6300 Å and OI 1356 Å observations, *J. Geophys. Res., 80*, 2327–2332, 1975.

Cleary, D. E., R. R. Meier, E. P. Gentieu, P. D. Feldman, and A. B. Christensen, An analysis of the effects of N_2 absorption on the O^+ 834 Å emission from rocket observations, *J. Geophys. Res., 94*, 17,281, 1989.

Daniell, Jr., R. E., D. J. Strickland, and J. R. Jasperse, Global monitoring of the ionosphere by optical techniques, in *Ionospheric Effects Symposium 1984: Effect of the Ionosphere on C^3I Systems*, Editors: J. M. Goodman, F. D. Clarke, J. A. Klobucher, and H. Soicher, Available from NTIS, 276–287, 1984.

Daniell, R. E., Jr., D. J. Strickland, D. T. Decker, J. R. Jasperse, and H. C. Carlson, Jr., Determination of ionospheric electron density profiles from satellite UV emission measurements. FY 84, *AFGL-TR-85-0099, Air Force Geophysics Laboratory*, 1985.

Daniell, R. E., Jr., D. T. Decker, D. N. Anderson, J. R. Jasperse, J. J. Sojka, and R. W. Schunk, A global ionospheric conductivity and electron density (ICED) model, in *Ionospheric Effects Symposium 1990: The Effect of the Ionosphere on Radiowave Signals and Systems Performance*, Editors: J. M. Goodman, J. A. Klobuchar, R. G. Joiner, and H. Soicher, Available from NTIS, 351–359, 1990.

Davies, K., *Ionospheric Radio*, Peter Peregrinus Ltd., 1989.

Decker, D. T., R. E. Daniell, Jr., J. R. Jasperse, and D. J. Strickland, Determination of ionospheric electron density profiles from satellite UV emission measurements, *Ultraviolet Technology*, R. E. Huffman, editor, *Proc. SPIE, 687*, 73–80, 1986.

Decker, D. T., R. E. Daniell, Jr., J. R. Jasperse, and D. J. Strickland, Determination of ionospheric electron density profiles from satellite UV emission measurements, in *Ionospheric Effects Symposium 1987: Effect of the Ionosphere on Communication, Navigation, and Surveillance Systems*, Editors: J. M. Goodman, J. A. Klobuchar, R. G. Joiner, and H. Soicher, Available from NTIS, 685–694, 1987.

Decker, D. T., J. M. Retterer, J. R. Jasperse, D. N. Anderson, R. W. Eastes, F. P. DelGreco, R. E. Huffman, and J. C. Foster, Determination of daytime midlatitude electron density profiles from satellite UV and in-situ data, *Ultraviolet Technology II*, R. E. Huffman, editor, *Proc. SPIE, 932*, 24–29, 1988.

Decker, D. T., J. M. Retterer, J. R. Jasperse, D. N. Anderson, F. J. Rich, and J. C. Foster, Determination of daytime midlatitude electron density profiles from limited real time measurements, in *Ionospheric Effects Symposium 1990: The Effect of the Ionosphere on Radiowave*

17.7. REFERENCES

Signals and Systems Performance, Editors: J. M. Goodman, J. A. Klobuchar, R. G. Joiner, and H. Soicher, Available from NTIS, 436–441, 1990.

DelGreco, F. P., R. W. Eastes, and R. E. Huffman, UV ionospheric remote sensing with the Polar BEAR satellite, *Ultraviolet Technology III*, R. E. Huffman, editor, *Proc. SPIE, 1158*, 46–50, 1989.

Fejer, B. G. and M. C. Kelley, Ionospheric irregularities, *Rev. Geophys. Spa. Phys., 18*, 401–454, 1980.

Fremouw, E. J., R. L Leadabrand, R. C. Livingston, M. D. Cousins, C. L. Rino, B. C. Jair, and R. A. Long, Early results from the DNA wideband experiment—Complex signal scintillation, *Radio Sci., 13*, 167, 1978.

Gerard, J. C., D. N. Anderson, and S. Matsushita, Magnetic storm effects on the tropical ultraviolet airglow, *J. Geophys. Res., 82*, 1977.

Heroux, L., M. Cohen, and J. E. Higgins, Electron densities between 110 and 300 km derived from solar EUV fluxes of August 23, 1972, *J. Geophys. Res., 79*, 5237–5244, 1974.

Heroux, L., M. Cohen, and J. E. Higgins, Improved calculations of electron densities between 110 and 300 km derived from solar EUV fluxes of August 23, 1972, *J. Geophys. Res., 80*, 4732–4734, 1975.

Huffman, R. E., F. J. LeBlanc, D. E. Paulsen, and J. C. Larrabee, Satellite vacuum ultraviolet airglow and auroral observations, *J. Geophys. Res., 85*, 2201–2215, 1980.

Huffman, R. E., D. E. Paulsen, F. J. LeBlanc, and J. C. Larrabee, Ionospheric and auroral measurements from space using vacuum ultraviolet emission, in *Ionospheric Effects Symposium 1981: Effect of the Ionosphere on Radiowave Systems*, Editors: J. M. Goodman, F. D. Clarke, and J. Aarons, Available from NTIS, 364–371, 1981.

Huffman, R. E., J. C. Larrabee, F. J. LeBlanc, and C. I. Meng, Ultraviolet remote sensing of the aurora and ionosphere for C^3I systems use, in *Ionospheric Effects Symposium 1984: Effect of the Ionosphere on C^3I Systems*, Editors: J. M. Goodman, F. D. Clarke, J. A. Klobucher, and H. Soicher, Available from NTIS, 288–294, 1984.

Huffman, R. E., J. C. Larrabee, F. J. LeBlanc, and C. I. Meng, Ultraviolet remote sensing of the aurora and ionosphere for C^3I systems use, *Radio Sci., 20*, 425–430, 1985.

Jasperse, J. R., Electron distribution function and ion concentrations in the Earth's lower ionosphere from Boltzmann-Fokker-Planck theory, *Planet. Space Sci., 25*, 743, 1977.

Jasperse, J. R., The photoelectron distribution function in the terrestrial ionosphere, *Physics of Space Plasmas, Proc. SPIE, 4*, 1981.

Kelley, M. C., *The Earth's Ionosphere, Plasma Physics and Electrodynamics*, Academic Press, 1989.

Kumar, S., S. Chakrabarti, F. Paresce, and S. Bowyer, The O^+ 834 Å dayglow: Satellite observations and interpretation with a radiation transfer model, *J. Geophys. Res., 88*, 9271–9279, 1983.

Leadabrand, R. L., M. J. Baron, J. Petriceks, and H. F. Bates, Chatanika, Alaska, auroral zone incoherent scatter facility, *Radio Sci., 7*, 747, 1972.

Link, R. and L. L. Cogger, A reexamination of the OI 6300 Å nightglow, *J. Geophys. Res., 93*, 9883–9892, 1988.

Lund, D. S., R. D. Hunsucker, H. F. Bates, and W. B. Murcray, Electron number densities in auroral irregularities: Comparison of backscatter and satellite data, *J. Geophys. Res., 72*, 1053–1059, 1967.

McCoy, R. P. and D. E. Anderson, Jr., Ultraviolet remote sensing of the F_2 ionosphere, in *Ionospheric Effects Symposium 1984: Effect of the Ionosphere on C^3I Systems*, Editors: J. M. Goodman, F. D. Clarke, J. A. Klobucher, and H. Soicher, Available from NTIS, 295–302, 1984.

McCoy, R. P., D. E. Anderson, Jr., and S. Chakrabarti, F_2 region ion densities from analysis of O^+ 834 Å airglow: A parametric study and comparisons with satellite data, *J. Geophys. Res., 90*, 12,257–12,264, 1985.

McCoy, R. P., K. D. Wolfram, R. R. Meier, L. J. Paxton, D. D. Cleary, D. K. Prinz, D. E. Anderson, Jr., A. B. Christensen, J. Pranke, G. G. Sivjee, and D. Kayser, Remote atmospheric and ionospheric detection

17.7. REFERENCES

system, *Ultraviolet Technology*, R. E. Huffman, editor, *Proc. SPIE*, 687, 142–149, 1986.

McCoy, R. P., L. J. Paxton, R. R. Meier, D. D. Cleary, D. K. Prinz, K. D. Wolfram, A. B. Christensen, J. B. Pranke, and D. C. Kayser, RAIDS: An orbiting observatory for ionospheric remote sensing from space, in *Ionospheric Effects Symposium 1987: Effect of the Ionosphere on Communication, Navigation, and Surveillance Systems*, Editors: J. M. Goodman, J. A. Klobucher, R. G. Joiner, and H. Soicher, Available from NTIS, 519–525, 1987.

Meier, R. R., Overview of UV remote sensing of the ionosphere, *NRL Memorandum Report 5292*, Naval Research Laboratory, 1984.

Meier, R. R., Ultraviolet spectroscopy and remote sensing of the upper atmosphere, *Space Sci. Rev.*, 58, 1–185, 1991.

Meier, R. R. and C. B. Opal, Tropical UV arcs: Comparison of brightness with f_oF_2, *J. Geophys. Res.*, 78, 3189–3193, 1973.

Meng, C. I., R. E. Huffman, R. A. Skrivanek, D. J. Strickland, and R. E. Daniell, Jr., Remote sensing of ionosphere by using UV and visible emissions, *Ultraviolet Technology*, R. E. Huffman, editor, *Proc. SPIE*, 687, 62–72, 1986.

Moore, J. G. and E. J. Weber, OI 6300 and 7774 Å airglow measurements of equatorial plasma depletions, *J. Atmos. Terr. Phys.*, 43, 651–658, 1981.

Morrison, R. A. and D. D. Cleary, A high resolution measurement of the photoabsorption cross section of N_2 near the 83.4 nm lines emitted by O^+, *EOS, Trans, Am. Geophys. U.*, 408, 1989.

Parks, G. K., *Physics of Space Plasmas, An Introduction*, Addison-Wesley, 1991.

Paxton, L. J. and D. J. Strickland, EUV imaging of the ionosphere from space, *Ultraviolet Technology II*, R. E. Huffman, editor, *Proc. SPIE*, 932, 190–202, 1988.

Rees, M. H., *Physics and Chemistry of the Upper Atmosphere*, Cambridge U. Press, 1989.

Rich, F. J. and Su. Basu, Ionospheric physics, Chapter 9 in *Handbook of Geophysics and the Space Environment*, A. S. Jursa, Scientific Editor, Air Force Geophysics Laboratory, NTIS Document Accession Number ADA 167000, 1985.

Richards, P. G. and D. G. Torr, A factor of 2 reduction in theoretical F2 peak electron density due to enhanced vibrational excitation of N_2 in summer at solar maximum, *J. Geophys. Res., 91*, 11,331–11,336, 1986.

Richards, P. G. and D. G. Torr, Ratios of photoelectron to EUV ionization rates for aeronomic studies, *J. Geophys. Res., 93*, 4060–4066, 1988.

Robinson, R. M. and R. R. Vondrak, Ultraviolet and x-ray remote sensing of high latitude ionization from space based platforms, in *Ionospheric Effects Symposium 1987: Effect of the Ionosphere on Communication, Navigation, and Surveillance Systems*, Editors: J. M. Goodman, J. A. Klobucher, R. G. Joiner, and H. Soicher, Available from NTIS, 695–704, 1987.

Roble, R. G., E. C. Ridley, and R. E. Dickinson, On the global mean structure of the thermosphere, *J. Geophys. Res., 92*, 8745–8758, 1987.

Rohrbaugh, R. P., W. B. Hanson, B. A. Tinsley, B. L. Cragin, J. P. McClure, and A. L. Broadfoot, Images of transequatorial bubbles based on field-aligned airglow observations from Haleakala in 1984–1986, *J. Geophys. Res., 94*, 6763–6770, 1989.

Sahai, Y., J. A. Bittencourt, H. Takahashi, and M. Mendillo, Comparison of a low-latitude ionospheric model with observations of OI 630 nm emission and ionospheric parameters, *Planet. Space Sci., 38*, 1243–1250, 1990.

Schunk, R. W., The terrestrial ionosphere, in *Solar-Terrestrial Physics*, R. L. Carovillano and J. M. Forbes, 609–676, Reidel, 1983.

Schunk, R. W. and J. J. Sojka, Approaches to ionospheric modelling, simulation, and prediction, *Adv. Space Res., 12*, 317–326, 1992.

Serafimov, K. B. and M. K. Serafimova, Thermospheric modelling based on satellite and ground-based measurments, *Adv. Space Res., 12*, 97–104, 1992.

17.7. REFERENCES

Sojka, J. J., Global scale, physical models of the F region ionosphere, *Rev. Geophys.*, *27*, 371–403, 1989.

Sojka, J. J., R. W. Schunk, J. D. Craven, L. A. Frank, J. Sharber, and J. D. Winningham, Modeled F region response to auroral dynamics based on Dynamics Explorer auroral observations, *J. Geophys. Res.*, *94*, 8993–9008, 1989.

Solomon, S. C. and V. J. Abreu, The 630 nm dayglow, *J. Geophys. Res.*, *94*, 6817-6824, 1989.

Strickland, D. J., Electron transport, chemistry, and optical emissions in the auroral E layer, *AFGL-TR-81-0042, Air Force Geophysics Laboratory*, ADA102345, 1981.

Strickland, D. J. and R. E. Daniell, Jr., UV emissions and the electron density in the auroral and low to mid-latitude daytime ionospheres, *AFGL-TR-82-0373, Air Force Geophysics Laboratory*, 1982.

Strickland, D. J., J. R. Jasperse,, and J. A. Whalen, Dependence of auroral FUV emissions of the incident electron spectrum and neutral atmosphere, *J. Geophys. Res.*, *88*, 8051–8062, 1983.

Strickland, D. J., R. E. Daniell, Jr., D. Decker, J. R. Jasperse, and H. C. Carlson, Determination of ionospheric electron density profiles from satellite UV emission measurements, *AFGL-TR-84-0140, Air Force Geophysics Laboratory*, 1984.

Strobel, D. F., T. R. Young, R. R. Meier, T. P. Coffey, and A. W. Ali, The night-time ionosphere: E-region and lower F-region, *J. Geophys. Res.*, *79*, 3171, 1974.

Strobel, D. F., C. B. Opal, and R. R. Meier, Photoionization rates in the nighttime E- and F-region ionosphere, *Planet. Spa. Sci.*, *28*, 1027–1033, 1980.

Szuszczewicz, E. P., Theoretical and experimental aspects of ionospheric structure: A global perspective on dynamics and irregularities, *Radio Sci.*, *21*, 351–362, 1986.

Szuszczewicz, E. P., P. Rodriquez, M. Singh, and S. Mango, Ionospheric irregularities and their potential impact on synthetic aperture radars, *Radio Sci.*, *18*, 765–774, 1983.

Tascione, T. F., H. W. Kroehl, R. Creiger, J. W. Freeman, R. A. Wolf, R. W. Spiro, R. V. Hilmer, J. W. Shade, and B. A. Hausman, New ionospheric and magnetospheric specification models, *Radio Sci.*, *23*, 211–222, 1988.

Tinsley, B. A., Field aligned airglow observations of transequatorial bubbles in the tropical F-region, *J. Atmos. Terr. Phys.*, *44*, 547–557, 1982.

Tinsley, B. A. and J. A. Bittencourt, Determination of F region height and peak electron density at night using airglow emissions from atomic oxygen, *J. Geophys. Res.*, *80*, 2333-2337, 1975.

Torr, M. R. and D. G. Torr, Ionization frequencies for Solar Cycle 21: Revised, *J. Geophys. Res.*, *90*, 6675–6678, 1985.

Vondrak, R. R., Incoherent scatter radar measurements of electric field and plasma in the auroral ionosphere, in *High Latitude Space Plasma Physics*, B. Hultqvist and T. Hagfors, editors, Plenum Press, 73–94, 1983.

Wadzinski, H. T. and J. R. Jasperse, Low energy electron and photon cross sections for O, N_2, O_2, and related data, *AFGL-TR-82-0008, Air Force Geophysics Laboratory*, 1982.

Weber, E. J., J. Aarons, and A. L. Johnson, Conjugate studies of an isolated equatorial irregularity region, *J. Geophys. Res.*, *88*, 3175–3180, 1983.

Weber, E. J., M. C. Kelley, J. O. Ballenthin, S. Basu, H. C. Carlson, J. R. Fleischman, D. A. Hardy, N. C. Maynard, R. F. Pfaff, P. Rodriguez, R. E. Sheehan, and M. Smiddy, Rocket measurements within a polar cap arc: Plasma, particle, and electric circuit parameters, *J. Geophys. Res.*, *94*, 6692–6712, 1989.

Index

Ångstrom, (Å), 7

ab initio electron density model, 283–287
absorption coefficient, 94
absorption cross section
   defined, 93
   measurements, 104–115
aeronomy, 3, 103
Airy disk, 30
airglow, 121–134
   AURIC code use of, 192
   day, 123–127
   defined, 121–123
   excitation processes, 121
   night, 127–134
   spectra,
     day, 127, 128,
     night, 129, 132, 133
   tables,
     EUV, 179
     MUV and FUV, 177
AKEBONO satellite (EXOS-D), 264
aperture, 26, 28, 36
Appleton anomaly, 131
array, 28, 47-48
ASSI/San Marcos satellite, 73, 122
atmospheric composition
   model number densities, 79, 80
   stratospheric sensing, 208–219
   upper atmospheric sensing, 227–245
   recommended methods, 243–244
atmospheric density, 244–245
atmospheric regions, 78, 82
atmospheric transmission, 12, 16, 17
   to/from aircraft, 58
   to/from balloon, 60
   to/from ground, 58
   to/from satellites, 62
   to/from shuttle, 62
   to/from sounding rocket, 62
atmospheric ultraviolet, 12, 17
atmospheric UV backgrounds, 171–186
   auroral levels, 182–183
   clutter in, 183–186
   EUV table, 179
   figure, 173
   limb maximum in MUV, 175
   LOWTRAN 7 values, 200–400 nm, 178
   MUV and FUV table, 177
   night limb maximum, 182
   relation to celestial, 171
   S3-4 total orbit scans, FUV,

179–181
  small comets and, 186
atmospheric window region, O_2,
    110–111
AURA satellite experiment, 287
AURIC UV code, 191–196
  current work on, 194–196
  figure illustrating, 193
aurora
  description, 137–150
  EUV image study, 267
  excitation of, 138, 140
  FUV images, 15, 146, 147,
    148, 263
  in-situ sensors for, 268
  satellite UV imagers,
    255–267
  satellite visible imagers,
    254–255
  spectra, 145
  tables of spectra
    EUV, 144
    MUV and FUV, 143
    NUV, 142
  x-ray image of, 269
auroral oval
  defined, 138, 140
  images, 13, 15, 146–148,
    263
  mapped, 139
  relation to K_p, 138
autoionization, 99–100

backgrounds, *see* atmospheric
    UV backgrounds
band, 24–25
bandwidth, 24–25
beacon experiment, satellite,
    292–293

Beer, 94
Bouguer, 94
branching ratio, 99
bromine, atomic (Br), 219
bromine monoxide (BrO), 218

calibration
  facility, 49–51
  laboratory, 49–51
  on-orbit, 51–52
Chapman layer, 94, 282
Chappius bands, 106
charge coupled device (CCD),
    28, 45–48
  intensified, 48
chlorine, atomic (Cl), 219
chlorine dioxide (OClO), 218
chlorine monoxide (ClO), 218
clutter, 183–185
comets, small, in DE-1 images,
    186
count equivalents, 25
counts (C), 25–28, 30–33
  background, 31, 33
  dark, 30–32
  miscellaneous, 32
critical frequency, 282
cross section
  absorption, 93
  atmospheric species values,
    104–115
  defined, 93
  photodissociation, 98–99
  photoionization, 99–100
  use described, 93–97
current, 25
curve-of-growth, 97

D-region, 82

INDEX

DE-1 satellite, 70
    auroral imaging, 255–257
    auroral modeling, 150
    imager description, 256–257
    ozone measurements, 216
    small comets, 186
deep UV, 9
Defense Meteorological Satellite Program, see DMSP satellite
diffraction gratings, 44–45
    holographic, 45
    mechanically ruled, 45
DMSP satellite, 253, 266, 268
    auroral imaging, 254–255
    Secondary Sensor Density experiment, 241
    UV sensors for, 295
Dobson unit for ozone, 211
Dynamics Explorer, see DE-1 satellite

E-region, 82
efficiency, (E), 26, 36
electromagnetic radiation, 7, 8
electron density
    83.4 nm model approach, 287–289
    ab initio model approach, 283–287
    and critical frequency, 282
    ionospheric layer values of, 82, 83
    radiowaves and, 280–282
    sounders for, 280–282
    UV radiance for, 283–291
    UV solar flux for, 291–292
electron Volt, 21–22
energy levels, 104, 105

environmental tests, 67–68
equatorial ionospheric emissions, 131–134
equatorial irregularities, 293–294
equivalent wavelengths, 104, 105
EUV (extreme ultraviolet), 9
exosphere, 78
extinction law, 93
extreme ultraviolet (EUV), 9

f_o, 282
f_oF_2, 282, 295
f-number, 28–29
F-region, 82
far ultraviolet (FUV), 8
field of view (FOV), (Ω), 27–29, 36, 37, 41, 47
fluorescence, 160–163
    atmospheric emitters, 161–163
    multiple scattering of, 160–161
focal length (f), 28, 29–30
fractional uncertainty, 31
FUV (far ultraviolet), 8, 10

geocorona, 11–12
geostationary, orbit, 65
global auroral imaging, 2, 253–270
Global ICED model, 297
global space weather systems, 2, 279, 294–297
    ionospheric models, 295–297
    satellite UV sensors, 295
global ultraviolet, 11–12
glow, spacecraft, 68–70
GOES satellite, 253
green line, oxygen

airglow, 122
  for auroral IBC definition, 141
  ground support equipment, 51

Hartley bands, 106
Herzberg continuum and bands, 109
HILAT satellite, 73
  auroral imaging from, 257–260
  beacon experiment on, 293
  program description, 257
helium (He)
  58.4 nm fluorescence, 163
  atmospheric sensing, 234
helium ion (He^+)
  30.4 nm fluorescence, 163
heterosphere, 82
homosphere, 78
Horizon Ultraviolet Program (HUP) shuttle experiment
  N/LBH 149.3 nm limb, 164
  N_2 LBH band limb, 238, 239
  O I 130.4 nm limb, 164
  O I 135.6 nm limb, 164, 241, 242
Huggins bands, 106
hydrogen atom (H)
  composition, 80
  Lyman α (121.6 nm)
    auroral emission, 149
    auroral imaging, 255, 265
    fluorescence, 162
    geocorona, 11, 233–234, 265
  measurement, 233
hydroxyl radical (OH)
  emission from fluorescence, 160–163
  measurement
    ground, 218
    space, MAHRS, 162

ICED ionospheric model, 296
imager, 45–47
imaging detectors, 47–49
  CCD, 48
  Codacon, 48
  Digicon, 47
  electrography, 48–49
  MAMA, 48
  Ranacon, 48
  Reticon, 47
  video tubes, 48
Imaging Spectroscopic Observatory, ISO, 73
infrared, 18–19
in-situ sensors, 213, 218, 236, 268
interface, 66
Interface Control Document, 66
interference filter, 39
International Reference Ionosphere, IRI, 296
International Ultraviolet Explorer (IUE), 48, 52
ionization cross sections
  defined, 99
  measurements, 104–115
ionosphere
  data bases, 296
  day midlatitude sensing 283–289
  description, 82
  disturbance, 292–294
  irregularities, 292–294

INDEX

models, 295–298
night sensing, 289–291
scintillation due to, 292
sounders, 280–282
irradiance, F, 24, 27
ISIS-II satellite imaging, 254

Joule, 21, 22

K, in radiance equations, 25–27, 36
KYOKKO satellite (EXOS-A), 255

Lambert, 94
LANDSAT satellite, 253
Loschmidt's number, 94
LOWTRAN codes, 190
LOWTRAN 7, 90, 190–191, 192
    validation, absorption, 197–198
    validation, MUV radiance, 197–199
LTE and UV, 200
Lyman α emission,
    see hydrogen atom
Lyman-Birge-Hopfield bands,
    see nitrogen, molecular

magnesium ion (Mg^+)
    fluorescence, 162
magnetic dip equator, 131
magnetospheric UV imaging, 270
Megabarn unit (Mb), 93
mesosphere, 78
microchannel plate (MCP), 45
micrometer, 7
micron, 7
middle ultraviolet (MUV), 8

minor species,
    composition, 77
    cross sections, 114–115
    stratospheric measurements, 218–219
MODTRAN code, 191
Molniya orbit, 64
MSIS model, 84
MUV (middle ultraviolet), 8

nanometer (nm), 7
near ultraviolet (NUV), 8
NEFD, 37
NER, 37
NIMBUS satellites, 216, 253
nitric oxide (NO)
    atmospheric measurement
        fluorescence method, 232–233
        from SME satellite, 233
    delta bands
        atmospheric composition using, 243
        night airglow emission, 130–131
    emission in AURIC, 195
    gamma bands
        atmospheric composition using, 243
        fluorescence, 162
        night airglow emission, 130–131
nitrogen, atomic (N),
    120 nm emission, 126
    149.3 nm emission, 126
    limb profile of, 164
nitrogen, atomic ion (N^+),
    214.3 nm emission, 126, 143
nitrogen dioxide (NO_2), 218

nitrogen, molecular (N_2),
   atmospheric measurment
      day airglow method,
         237–240
      EUV absorption method,
         230–232
   Birge-Hopfield bands (BH),
      124–125
   energy thresholds, 105
   Lyman-Birge-Hopfield
      bands (LBH)
      airglow emission, 123
      atmospheric composition
         using, 236–240
      auroral emission model,
         149–150
      auroral image using, 147,
         256, 260, 264, 265
      electron densities with,
         283–287
      in spacecraft glow, 68–70
      limb profiles, 238, 239
      photon cross sections, 109,
         111–113
   Second positive bands (2P),
      124
      atmospheric composition
         using, 240
   Vegard-Kaplan bands (VK),
      124
nitrogen, molecular ion (N_2^+),
   auroral emission calculation,
      149–150, 151
   auroral emission image, 148
   fluorescence, 161
   from ground, 149
noctilucent clouds (NLC), 163
NUV (near ultraviolet), 8

OGO-4 satellite, 72
   airglow measurements, 122
   auroral measurements, 141,
      254
   equatorial airglow, 131
   ozone measurements, 216
   Rayleigh scatter, 156
optical density unit, 95
optical depth, 95
Orbiting Geophysical
   Observatory,
      see OGO-4 satellite
OSO-III satellite data, 230
oxygen, atomic (O)
   130.4 nm multiplet
      airglow emission, 125
      auroral emission model,
         149–150, 151
      auroral image in, 146,
         256, 260, 264, 265
      auroral spectra, 143, 145
      equatorial emission,
         131–134
      limb profile, 164
      solar scatter, 235
   135.6 nm multiplet
      airglow emission, 125
      atmospheric composition
         using, 240–242
      auroral image in, 256,
         260, 264, 265
      auroral oval image of, 15
      electron densities with,
         284
      equatorial emission from,
         131–134
      limb profile, 164
   atmospheric measurement
      day airglow approach,

INDEX 313

    240–241
    EUV approach, 230–232
    fluorescence approach,
      234–236
    in-situ approach, 236
    energy thresholds, 105
    fluorescence, 162
    photoionization continuum,
      113–114
    photon cross sections, 105,
      113–114
    resonance lines, 113
oxygen, atomic ion, (O^+),
    83.4 nm multiplet, 125–126
    electron densities with,
      287–289
    fluorescence, 163
oxygen, molecular
    atmospheric measurement
      with EUV absorption,
        230–232
      with FUV absorption,
        229–230
    atmospheric window region,
      110–111
    energy thresholds, 105
    Herzberg bands, 109
      atmospheric composition
        using, 243
      limb emission, 182
      night airglow, 127–130
    Herzberg continuum, 109
    photoionization continuum,
      111
    photon cross sections, 107,
      109–111
    Schumann-Runge bands,
      109–110
      in AURIC code, 195

    Schumann-Runge
      continuum, 110
ozone, 2, 207–218
    Chappius bands, 106, 211
    Dobson unit defined, 211
    energy thresholds, 105
    formation, 210
    global zonal distribution,
      211, 212
    Hartley bands, 106, 211
    hole, 12
    Huggins bands, 106, 211
    in-situ measurements of
      reactants, 219
    photon cross sections,
      106–107
    satellite measurements
      DE-1, 216
      Meteor-3, 217
      Nimbus, 216
      NOAA, 216
      OGO-4, 216
      SAGE I and II, 215
      SME, 216
      UVSP, 215
    stratospheric, global
      backscatter approach,
        156, 215–218
      data bases, 218
      occultation approach,
        214–215
      profile, 217
      total ozone, 217
    stratospheric, local
      LIDAR methods, 214
      passive absorption, 213
    TOMS sensor, 216–217
    tropospheric, 209

P78-1 satellite, 72
    airglow measurements, 122
    atmospheric composition measurements, 241
    auroral measurements, 141, 255
    EUV airglow, 127, 131
    EUV aurora, 141, 144
PCTRAN 7 code, 191
phenomenology, defined, 192
photoabsorption, 22, 93–97
photocathode, 38–39
    cesium iodide, 38, 43
    cesium telluride, 38, 43
    EUV, 39
    potassium bromide, 38
    trialkali, 38
photodiode, 39
photodissociation, 22, 97–99
    cross section, 98
    defined, 97
    yield, 98
photoemission, 38
photoionization, 22, 97–100
    cross section, 99
        measurements, 104–115
    defined, 97
    yield, 99
photoionization continuum,
    nitrogen, molecular, 105, 112–113
    oxygen, atomic, 105, 113–114
    oxygen, molecular, 105, 111
photon, 21
    counting, 25
    energy equivalent, 22
photometer, 37–40
photomultiplier, 38

plume, rocket exhaust, 70–72
pixel, 28, 47
Planck's constant, 21
Planck's relation, 21
Polar BEAR satellite, 73
    auroral imager, 260–263
    auroral oval image, 15
    beacon experiment, 292–293
    program description, 257–258
    simultaneous auroral images, 145, 146–148
polar cap structure, 11
polar irregularities, 294
polar orbit, satellite, 63–64
polar mesospheric clouds (PMC), 163–167
    and global change, 167
    identification of, 163
    limb profile of, 166
POLAR satellite, 266
predissociation, 99
preionization, 100
PRISM model, 297

quantum, 21
quantum yield, 98

radar, incoherent scatter, 281
radiance, B, 24, 27, 36
radiance and transmission codes, 189–201
radiant flux, 22
radiant intensity, J, 24, 25–27
radiation hardening, 97
radiometer, 37
radiometric equations, 25–29
radiometry, 21–33

INDEX

RAIDS experiment, 288
Rayleigh scattering, 93, 155–160
   cross section for, 156
   experimental measurement,
      156–157
   limb calculations, 158–160
   theoretical calculation, 157
   volcano emission and,
      160, 207
Rayleigh unit, 22–23
red line, oxygen, 122
resonance lines,
   fluorescence from, 160

S3-4 satellite, 72
   airglow measurements, 122
   auroral measurements, 145,
      255
   day airglow spectra, 128
   LOWTRAN 7 comparison,
      197, 199
   night airglow spectra, 129
   night zonal spectra, 132,
      133
   photometer total orbit
      scans, 179–181
   Rayleigh scatter, 157
SAGE I and II satellites, 215
satellite drag, for density,
   244–245
satellites, Earth orbiting, 62–66
scene generator, 200
Schumann-Runge bands,
   109–110
Schumann-Runge continuum,
   110
scintillation experiment,
   292–293
sensitivity, (S_B), 36–37

imager, 47
photometer, 40
spectrometer, 41
SHARC code, 200
shuttle, see Space Transportation
   System
signal-to-background, 25, 33
signal-to-noise, 25, 32–33
SMAART code, 200
small comets, DE-1, 186
SME satellite, 72
   description, 165, 167
   nitric oxide measurement,
      233
   ozone measurement, 216
   PMC, 163, 165, 166
solar blind, 9, 38
solar constant, 92
solar EUV
   for density modeling,
      244–245
   variability, 92
solar flux,
   quiet sun, 88–91
   variability, 92
Solar Maximum Mission, 215
Solar Mesosphere Explorer,
   see SME satellite
solar reference spectrum, 88
solar zenith angle (SZA), 17, 94
solid propellant,
   as emission source, 71
sounders, ionospheric, 280–282
space telescope, Hubble, 47
Space Transportation System
   (shuttle)
   atmospheric remote sensing
      from, 62
   HUP data from, 164, 238,

239, 241, 242
spectrograph, 41
spectrometer, 40–44
    Ebert-Fastie, 40, 44
    Rowland circle, 44
    Wadsworth, 44
speed of light (c), 21
SPOT satellite, 253
stabilization, satellite
    inertial, 65
    spin, 65
    three-axis, 65–66
standard
    atmospheres, 77, 79 80, 82, 84
    detector, 49
    flight sources, 51–52
    light source, 49, 51
    synchrotron source, 49
    temperature and pressure (STP), 77, 94
stars,
    UV, as standards, 52
STP (standard temperature and pressure), 77, 94
stratosphere, 78
Stratospheric Aerosol and Gas Experiment, *see* SAGE I and II satellites
SZA (solar zenith angle), 17

technology transition, 3, 218
test bed, 200
thermosphere, 78
throughput (optical), 27–28, 41
thruster, cold gas, 71
TIGCM model, 84
topside ionosphere, 82
topside sounder, 281–282

transmission
    atmospheric, 12, 16, 17, 57–62, 95
    codes, 189–201
    defined, 95
tropical UV airglow, 11, 12, 131–134
troposphere, 78, 207–210

ultraviolet (UV), 7
    advantages, 18
    atmospheric, 12, 14–17
    deep, 9
    defined, 7
    disadvantages, 18–19
    extreme (EUV), 9
    far (FUV), 8
    global, 11–12
    middle (MUV), 8
    near (NUV), 8
    solar-blind, 9, 11
    vacuum (VUV), 8
    windowless, 11
unit optical depth, 12, 16, 95
UV-A, 9
UV-B, 9

vacuum ultraviolet (VUV), 8
VIKING satellite, 73
    auroral imager, 263–264
volcanic emission seen with UV, 160, 207–208
VUV (vacuum ultraviolet), 8

watt, 21–22, 25, 26
wavelength band, 24–25
wavelength interval, 24–25
wavenumber (cm^{-1}), 21
Wideband satellite, 292–293

INDEX

window
    magnesium fluoride, 38
    lithium fluoride, 41
windowless ultraviolet, 11

x-ray auroral imagers, 267–268
XUV (extreme ultraviolet), 9

International Geophysics Series

EDITED BY

RENATA DMOWSKA

Division of Applied Sciences
Harvard University
Cambridge, Massachusetts

JAMES R. HOLTON

Department of Atmospheric Sciences
University of Washington
Seattle, Washington

Volume 1   BENO GUTENBERG. Physics of the Earth's Interior. 1959*

Volume 2   JOSEPH W. CHAMBERLAIN. Physics of the Aurora and Airglow. 1961*

Volume 3   S.K. RUNCORN (ed.). Continental Drift. 1962*

Volume 4   C. E. JUNGE. Air Chemistry and Radioactivity. 1963*

Volume 5   ROBERT G. FLEAGLE AND JOOST A. BUSINGER. An Introduction to Atmospheric Physics. 1963*

Volume 6   L. DUFOUR AND R. DEFAY. Thermodynamics of Clouds. 1963*

Volume 7   H. U. ROLL. Physics of the Marine Atmosphere. 1965*

Volume 8   RICHARD A. CRAIG. The Upper Atmosphere: Meteorology and Physics. 1965*

Volume 9   WILLIS L. WEBB. Structure of the Stratosphere and Mesosphere. 1966*

Volume 10   MICHELE CAPUTO. The Gravity Field of the Earth from Classical and Modern Methods. 1967*

Volume 11   S. MATSUSHITA AND WALLACE H. CAMPBELL (eds.) Physics of Geomagnetic Phenomena. (In two volumes.) 1967*

Volume 12   K. YA. KONDRATYEV. Radiation in the Atmosphere. 1969*

*Out of print.

INTERNATIONAL GEOPHYSICS SERIES

Volume 13 E. PALMEN AND C. W. NEWTON. Atmospheric Circulation Systems: Their Structure and Physical Interpretation. 1969

Volume 14 HENRY RISHBETH AND OWEN K. GARRIOTT. Introduction to Ionospheric Physics. 1969*

Volume 15 C. S. RAMAGE. Monson Meteorology. 1971*

Volume 16 JAMES R. HOLTON. An Introduction to Dynamic Meteorology. 1972*

Volume 17 K. C. YEH AND C. H. LIU. Theory of Ionospheric Waves. 1972*

Volume 18 M. I. BUDYKO. Climate and Life. 1974*

Volume 19 MELVIN E. STERN. Ocean Circulation Physics. 1975

Volume 20 J. A. JACOBS. The Earth's Core. 1975*

Volume 21 DAVID H. MILLER. Water at the Surface of the Earth: An Introduction to Ecosystem Hydrodynamics. 1977

Volume 22 JOSEPH W. CHAMBERLAIN. Theory of Planetary Atmospheres: An Introduction to Their Physics and Chemistry. 1978*

Volume 23 JAMES R. HOLTON. An Introduction to Dynamic Meteorology, Second Edition. 1979*

Volume 24 ARNETT S. DENNIS. Weather Modification by Cloud Seeding. 1980

Volume 25 ROBERT G. FLEAGLE AND JOOST A. BUSINGER. An Introduction to Atmospheric Physics, Second Edition. 1980

Volume 26 KUO-NAN LIOU. An Introduction to Atmospheric Radiation. 1980

Volume 27 DAVID H. MILLER. Energy at the Surface of the Earth: An Introduction to the Energetics of Ecosystems. 1981

Volume 28 HELMUT E. LANDSBERG. The Urban Climate. 1981

Volume 29 M. I. BUDYKO. The Earth's Climate: Past and Future. 1982

Volume 30 ADRIAN E. GILL. Atmosphere – Ocean Dynamics. 1982

Volume 31 PAOLO LANZANO. Deformations of an Elastic Earth. 1982*

Volume 32 RONALD T. MERRILL AND MICHAEL W. MCELHINNY. The Earth's Magnetic Field: Its History, Origin, and Planetary Perspective. 1983

Volume 33 JOHN S. LEWIS AND RONALD G. PRINN. Planets and Their Atmospheres: Origin and Evolution. 1983

Volume 34 ROLF MEISSNER. The Continental Crust: A Geophysical Approach. 1986

Volume 35 M. U. SAGITOV, B. BODRI, V. S. NAZARENKO, AND KH. G. TADZHIDINOV. Lunar Gravimetry. 1986

Volume 36 JOSEPH W. CHAMBERLAIN AND DONALD M. HUNTEN. Theory of Planetary Atmospheres: An Introduction to Their Physics and Chemistry, Second Edition. 1987

Volume 37 J. A. JACOBS. The Earth's Core, Second Edition. 1987

Volume 38 J. R. APEL. Principles of Ocean Physics. 1987

Volume 39 MARTIN A. UMAN. The Lightning Discharge. 1987

Volume 40 DAVID G. ANDREWS, JAMES R. HOLTON, AND CONWAY B. LEOVY. Middle Atmosphere Dynamics. 1987

Volume 41 PETER WARNECK. Chemistry of the Natural Atmosphere. 1988

Volume 42 S. PAL ARYA. Introduction to Micrometeorology. 1988

Volume 43 MICHAEL C. KELLEY. The Earth's Ionosphere. 1989

Volume 44 WILLIAM R. COTTON AND RICHARD A. ANTHES. Clouds and Precipitating Storms. 1989

Volume 45 WILLIAM MENKE. Geophysical Data Analysis: Discrete Inverse Theory, Revised Edition. 1989

Volume 46 S. GEORGE PHILANDER. El Niño, La Niña, and the Southern Oscillation. 1990

Volume 47 ROBERT A. BROWN. Fluid Mechanics of the Atmosphere. 1991

Volume 48 JAMES R. HOLTON. An Introduction to Dynamic Meteorology, Third Edition. 1992

Volume 49 ALEXANDER A. KAUFMAN. Geophysical Field Theory and Method. Part A: Gravitational, Electric, and Magnetic Fields. 1992

Volume 50 SAMUEL S. BUTCHER AND ROBERT CHARLSON. Global Biogeochemical Cycles. 1992

Volume 51 BRIAN EVANS AND TENG-FONG WONG. Fault Mechanics and Transport Properties in Rock. 1992

Volume 52 ROBERT E. HUFFMAN. Atmospheric Ultraviolet Remote Sensing. 1992